AF580870

**LACAME 2004**

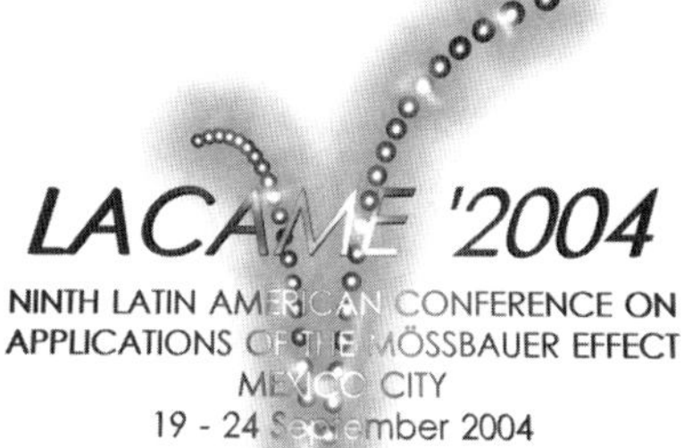

# LACAME 2004

*Proceedings of the 9th Latin American Conference on the Applications of the Mössbauer Effect (LACAME 2004) held in Mexico City, Mexico, 19-24 September 2004*

*Edited by*


R. C. MERCADER

*Universidad Nacional de La Plata, La Plata, Argentina*

J. R. GANCEDO

*Instituto de Quimica Fisica Rocasolano, Madrid, Spain*

A. CABRAL PRIETO

*Instituto Nacional de Investigaciones Nucleares, Mexico City, Mexico*

and

E. BAGGIO-SAITOVITCH

*Centro Brasileiro de Pesquisas Físicas, Rio de Janeiro, Brazil*



Reprinted from *Hyperfine Interactions*
Volume 161, Nos. 1–4 (2005)

**A C.I.P. Catalogue record for this book is available from the Library of Congress.**

ISBN 3-540-28959-3

---

Published by Springer
P.O. Box 990, 3300 AZ Dordrecht, The Netherlands

Sold and distributed in North, Central and South America
by Springer
101 Philip Drive, Norwell, MA 02061, U.S.A.

In all other countries, sold and distributed
by Springer
P.O. Box 990, 3300 AZ Dordrecht, The Netherlands

*Printed on acid-free paper*



Printed in The Netherlands

# Table of Contents

Hyperfine Interactions (2005) 161:1–2
DOI 10.1007/s10751-005-9185-1


# Ninth Latin American Conference on the Applications of the Mössbauer Effect, LACAME 2004 Mexico City, Mexico, 19–24 September 2004

**Organizing Committee**

| | |
|---|---|
| Noel Nava (Chair) | *Instituto Mexicano del Petróleo* |
| Humberto Arriola | *Universidad Nacional Autónoma de México* |
| Agustín Cabral | *Instituto Nacional de Investigaciones Nucleares* |
| Jesús Soberón | *Secretaría de Educación Pública* |
| Arturo García | *Instituto Politécnico Nacional* |
| Fabiola Monroy | *Instituto Nacional de Investigaciones Nucleares* |
| Leticia Aguilar | *Secretaría de Educación Pública* |
| Germán Vázquez | *Instituto Mexicano del Petróleo* |

**Latin American Committee**

| | |
|---|---|
| E. Baggio-Saitovitch | *Brasil* |
| N. R. Furet Bridón | *Cuba* |
| F. González Jiménez | *Venezuela* |
| J. A. Jaén | *Panama* |
| R. C. Mercader | *Argentina* |
| N. Nava | *Mexico* |
| V. A. Peña Rodríguez | *Peru* |
| G. A. Pérez Alcázar | *Colombia* |

SPONSORS

Universidad Nacional Autónoma de México, UNAM
Secretaría de Educación Pública, SEP, México
Instituto Nacional de Investigaciones Nucleares, ININ, México
Centro Latinoamericano de Física, CLAF, Rio de Janeiro
Abdus Salam International Centre for Theoretical Physics, ICTP, Trieste
Consejo Nacional para la Cultura y las Artes, CONACULTA-INBA, México
Museo Nacional de Arte, MUNAL, México
International Atomic Energy Agency, IAEA, Vienna
United Nations Educational, Scientific and Cultural Organization, UNESCO
Universidad Autónoma Metropolitana-Ixtapalapa, UAMI, México
Patronato del Museo Nacional de Arte, A. C., PMNA AC, México

# Preface

The Mössbauer Spectroscopy, which is a technique that yields unique information both at the atomic and structural levels, can be applied to a wide range of fields and has a simple and not too expensive experimental setup. Because of these distinctive features, the Mössbauer laboratories have been feasible agents for the scientific growth in many developing countries. In Latin America, since the first conference held in Rio de Janeiro in 1988, the role of regional meetings has been instrumental in the development of Mössbauer spectroscopy, and hence of scientific and technological research. Since then, the Latin American Mössbauer community has gained momentum, scientists have been attracted to the field and the number of laboratories across the region has increased.

In September 2004, the ninth Latin American Conference on the Applications of the Mössbauer Effect, LACAME 2004, was held in Mexico City. It followed the series of conferences that have been hosted successfully each two years in Rio de Janeiro, Havana, Buenos Aires, Santiago de Chile, Cusco, Cartagena de Indias, Caracas, and Panama City. These meetings have kept their original aim: to gather Mössbauer spectroscopists of Latin America to discuss their interests related to the applications of the Mössbauer effect, foster collaborations, interchange experiences, and give and receive help.

The vast geography of Latin America causes that the scientists of this region are to a certain extent isolated from the main scientific centers. In particular the young researchers and students only can afford attending the regional meetings. Quite a few scientists from developed countries have found interesting to hold scientific links with Latin American colleagues. They attend these meetings often providing new ways of collaboration in addition to offering their expertise and knowledge about the recent advances of the technique.

The Mexico meeting has developed in the typical friendly and pleasant atmosphere common to the LACAMEs that has usually helped finding opportunities for informal scientific exchange among the participants. We hope that this LACAME will also further help the scientific development in the region.

*R. C. Mercader, La Plata, Argentina*
*J. R. Gancedo, Madrid, Spain*
*A. Cabral Prieto, Mexico*
*E. Baggio-Saitovitch, Rio de Janeiro, Brasil*

Hyperfine Interactions (2005) 161:3–9
DOI 10.1007/s10751-005-9186-0


# Effect of the Calcination Atmosphere on the Structural Properties of the Reduced Fe/SiO$_2$ System

A. M. ALVAREZ[1], J. F. BENGOA[1], M. V. CAGNOLI[1], N. G. GALLEGOS[1], S. G. MARCHETTI[1] and R. C. MERCADER[2,*]
[1]*CINDECA, CONICET, CICPBA, Facultad de Ciencias Exactas, Facultad de Ingeniería, Universidad Nacional de La Plata, 47 N° 257, 1900, La Plata, Argentina; e-mail: anamalvz@quimica.unlp.edu.ar*
[2]*Departamento de Física, IFLP, Facultad de Ciencias Exactas, Universidad Nacional de La Plata, 1900, La Plata, Argentina; e-mail: mercader@fisica.unlp.edu.ar*

**Abstract.** Iron supported systems are frequently used as catalysts in the Fischer–Tropsch synthesis being the $Fe^0$ the active phase for the reaction. We have studied the influence of the calcination atmosphere (air or nitrogen) on the iron oxide reducibility and the metallic iron particle size obtained in Fe/SiO$_2$ system. We have impregnated a silicagel with $Fe(NO_3)_3 \cdot 9H_2O$ aqueous solution and the solid obtained was calcinated in air or $N_2$ stream. These precursors, with 5% (wt/wt) of Fe, were characterized by Mössbauer Spectroscopy at 298 and 15 K. Amorphous $Fe_2O_3$ species with 3 nm diameter in the former, and $\alpha$-$Fe_2O_3$ crystals of 48 nm diameter were detected in the last one. Both precursors were reduced in $H_2$ stream. Two catalysts were obtained and characterized by Mössbauer spectroscopy in controlled atmosphere at 298 and 15 K, CO chemisorption and volumetric oxidation. $\alpha$-$Fe^0$, $Fe_3O_4$ and $Fe^{2+}$ were identified in the catalyst calcined in air. Instead, only $\alpha$-$Fe^0$ was detected in the catalyst calcined in $N_2$. The iron metallic crystal sizes were estimated as ≈2 nm for the former and ≈29 nm for the last one. The different oxide crystal sizes, obtained from the diverse calcination atmospheres, have led to different structural properties of the reduced solids. It has been possible to reduce totally the existing iron in an Fe/SiO$_2$ system with iron loading lower than 10% (wt/wt).

## 1. Introduction

Iron-supported systems are frequently used as catalysts in the Fischer–Tropsch reaction (hydrocarbon synthesis by CO hydrogenation). The physical–chemical properties of these catalytic solids, like the percentage of metallic iron at the surface, crystallite size, degree of reducibility of the precursor Fe oxides to $Fe^0$, etc., have strong influence on their activity and selectivity. In these solids, the active species is $Fe^0$. To get the highest $Fe^0$ content it is very important to control the oxide-support interaction because the reducibility degree significantly depends on it.

* Author for correspondence.

In this kind of systems with very small iron particles, low iron content and metallic particles highly prone to re-oxidation, traditional methods like X-ray diffraction (XRD) and transmission electron microscopy (TEM) are not useful because the techniques cannot be applied '*in situ*' on the reduced sample. Instead, a powerful approach to study these systems is to apply Mössbauer spectroscopy (MS) in controlled atmosphere and low temperature.

In previous studies on the $SiO_2$ supported Fe system, we have demonstrated that the control of the process variables of the iron salt decomposition by calcinations is a key step to reach the highest $Fe^0$ content [1, 2]. However, some aspects deemed a deeper understanding. In this work we study the influence of the calcination atmosphere (air or nitrogen) on the iron oxide reducibility and the metallic iron crystallite size obtained when iron oxide is supported on silica.

## 2. Experimental

We have impregnated a silicagel of 400 $m^2/g$ specific surface area with $Fe(NO_3)_3.9H_2O$ aqueous solution by incipient wetness method. The solid obtained was divided in two fractions, and they were calcinated at 698 K during 8 h in air (*p*-$Fe/SiO_2$(A)) or $N_2$ stream (*p*-$Fe/SiO_2$(N)). These precursors, with 5.5% (wt/wt) of Fe, determined by atomic absorption spectroscopy, were characterized by $N_2$ adsorption and Mössbauer spectroscopy at room temperature (RT) and 15K. Both precursors were reduced in $H_2$ stream at 698 K during 26 h obtaining *c*-$Fe/SiO_2$(A) and *c*-$Fe/SiO_2$(N). They were characterized by Mössbauer spectroscopy in controlled atmosphere at 298 and 15 K, CO chemisorption and volumetric oxidation.

Mössbauer spectra were recorded with a standard 512 channels spectrometer with transmission geometry. Samples were placed in a helium closed-cycle refrigerator at temperatures ranging from 15 to 298 K. A source of $^{57}Co$ in Rh matrix of nominally 50 mCi, at room temperature, was used. Velocity calibration was performed against a 12 μm thick $\alpha$-Fe foil. All isomer shifts are referred to this standard at 298 K. The spectra were fitted to Lorentzian line-shapes by a least-squares non-linear with constraints computer code. Lorentzian lines were considered for each spectrum component.

## 3. Results and discussion

The results of the $N_2$ adsorption techniques showed that the maxima of the bimodal pore diameter distribution [3] of silica changed from 10 to 6 nm and from 16 to 11 nm after impregnation and calcination in air (*p*-$Fe/SiO_2$(A)). Instead, in *p*-$Fe/SiO_2$(N) the same pore distribution as the support was observed. These results suggest that the iron species might be located inside the pores only in the first sample.

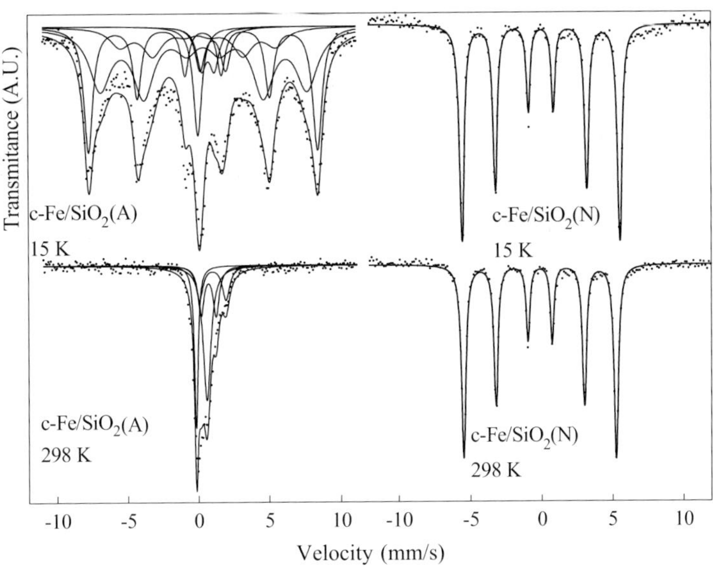


*Figure 1.* Mössbauer spectra of *c*-$Fe/SiO_2$(A) and *c*-$Fe/SiO_2$(N) at 298 and 15 K.

The existing iron species were identified using Mössbauer spectroscopy between 298 and 15 K. In previous works [1, 2] we have stated that the only iron species in both solids was $\alpha$-$Fe_2O_3$. However, in a more detailed and recent study, we have come to the conclusion that in *p*-$Fe/SiO_2$(A) the iron species is instead amorphous $Fe_2O_3$ [3].

Assuming valid the Néel–Brown model [4], we have estimated that the iron oxide particle size is $\approx$ 3 nm diameter in *p*-$Fe/SiO_2$(A). Applying the collective magnetic excitations model [4] in *p*-$Fe/SiO_2$(N), the estimated $\alpha$-$Fe_2O_3$ crystal diameter is about 48 nm. These findings indicate that, since the average pore diameter of silica is lower than 16 nm, in *p*-$Fe/SiO_2$(N), the iron oxide particles must be located outside the pores. Instead, in *p*-$Fe/SiO_2$(A), the particle sizes are lower than the pore diameter, allowing a possible location inside the pores. In connection with the $N_2$ adsorption results, we can conclude that in *p*-$Fe/SiO_2$(A) the iron particles are mainly inside the pores, while in *p*-$Fe/SiO_2$(N) are totally outside.

The Mössbauer spectra of both reduced solids in controlled $H_2$ atmosphere at 298 and 15 K are shown in Figure 1. The spectrum of *c*-$Fe/SiO_2$(A) at 298 K showed a central signal with three shoulders at the positive velocity range. Two doublets and two singlets were used in the fitting. When the temperature decreased to 15 K, a complex spectrum can be observed: a sextet with asymmetric internal sides and an important central signal. Three sextets, two doublets and one singlet were necessary to fit it. Considering the hyperfine parameters (Table I) obtained in fitting at 298 K, the signals were assigned to $Fe^{2+}$ [5], superparamagnetic $Fe^0$ ($Fe^0$(sp)) [6] and $Fe_3O_4$ (sp) [7] and, at 15 K, to $Fe^{2+}$, $Fe^0$(sp), magnetic $\alpha$-$Fe^0$ ($Fe^0$ (m)) and $Fe_3O_4$ (m) [8].

*Table I.* Mössbauer parameters of $c$-Fe/SiO$_2$(A) and $c$-Fe/SiO$_2$(N) at 298 and 15 K

| Sample | Parameters | $c$-Fe/SiO$_2$(A) | | $c$-Fe/SiO$_2$(N) | |
|---|---|---|---|---|---|
| Species | | 15 K | 298 K | 15 K | 298 K |
| $Fe_3O_4$(m) | $H$ (kG) | 502 ± 1 | – | – | – |
| | $\delta$(Fe) (mm/s) | 0.49 ± 0.01 | – | – | – |
| | 2ε (mm/s) | −0.05 ± 0.03 | – | – | – |
| | % | 28 ± 2 | – | – | – |
| $Fe_3O_4$(m) | $H$ (kG) | 454 ± 5 | – | – | – |
| | $\delta$(Fe) (mm/s) | 0.53 ± 0.03 | – | – | – |
| | 2ε (mm/s) | 0.00 (+) | – | – | – |
| | % | 41 ± 5 | – | – | – |
| α-Fe(m) | $H$ (kG) | 342 (+) | – | 345 ± 1 | 333 ± 1 |
| | $\delta$(Fe) (mm/s) | 0.13 (+) | – | 0.12 ± 0.01 | 0.01 ± 0.01 |
| | 2ε (mm/s) | 0.00 (+) | – | −0.01 ± 0.01 | −0.02 ± 0.01 |
| | % | 14 ± 3 | – | 100 ± 3 | 100 ± 3 |
| $Fe^{2+}$ (octahedral) | Δ (mm/s) | 1.80 (+) | 1.80 ± 0.04 | – | – |
| | $\delta$(Fe) (mm/s) | 1.21 (+) | 1.08 ± 0.03 | – | – |
| | % | 6 ± 1 | 27 ± 4 | – | – |
| $Fe^{2+}$ (tetrahedral) | Δ (mm/s) | 0.90(+) | 0.88 ± 0.07 | – | – |
| | $\delta$(Fe) (mm/s) | 0.86(+) | 0.83 ± 0.03 | – | – |
| | % | 5 ± 1 | 26 ± 7 | – | – |
| $Fe^0$(sp) | $\delta$(Fe) (mm/s) | 0.13 (+) | −0.06 ± 0.01 | – | – |
| | % | 6 ± 1 | 25 ± 3 | – | – |
| $Fe_3O_4$(sp) | $\delta$(Fe) (mm/s) | – | 0.82 ± 0.02 | – | – |
| | % | – | 22 ± 4 | – | – |

(+) parameter fixed in fitting.
$H$: hyperfine magnetic field.
$\Delta$: quadrupole splitting.
$\delta$: isomer shift referred to α-Fe at 298 K.
2ε: quadrupole shift.

The existence of $Fe^0$ (sp) at 298 K was confirmed because at 15 K this subspectrum splits into a magnetic sextet with typical hyperfine parameters of α-$Fe^0$ (m) with very broad peaks. This result is in agreement with that obtained by Bødker *et al.* [6] for iron crystals of about 4 nm in an Fe/C system. Two characteristic sextets of $Fe_3O_4$ [8] were observed at 15 K. Mørup *et al.* [9] found that $Fe_3O_4$ particles of about 6 nm show a magnetic signal at low temperatures, that collapses to a singlet at 227 K.

Taking into account these results, we assigned the singlet with isomer shift 0.79 ± 0.03 mm/s to $Fe_3O_4$ (sp) obtained in $c$-Fe/SiO$_2$(A) at 298 K. On the other hand, Lehlooh *et al.* [10] found that $Fe_3O_4$ particles of 3 nm display a well-defined doublet at room temperature instead of the singlet mentioned above. Therefore, the doublets observed at 298 K could be assigned to $Fe_3O_4$ (sp) with a size lower than the fraction that produces the singlet. However, Clausen *et al.* [5]

*Table II.* CO chemisorption values, experimental and calculated $O_2$ uptakes and average volumetric-superficial diameter of the $Fe^0$ crystals in $c$-$Fe/SiO_2$(A) and $c$-$Fe/SiO_2$(N)

| Sample | Experimental $O_2$ uptake (μmol/g of solid) | Calculated $O_2$ uptake (μmol/g of solid) | CO chemisorption (μmol/g of solid) | $d_{VA}$ (Å) |
|---|---|---|---|---|
| $c$-$Fe/SiO_2$(A) | 242 ± 12 | 241 ± 27 | 35 ± 2 | 22 |
| $c$-$Fe/SiO_2$(N) | 702 ± 35 | 725 ± 25 | 13 ± 1 | 287 |

assigned to $Fe^{2+}$ located in tetrahedral and octahedral surface sites of silica, two doublets with similar parameters to those found by us. Therefore, at 298 K it is not possible to discern between an $Fe^{2+}$ species and very small $Fe_3O_4$ particles. The spectrum at 15 K shows:

- a relative decrease of both doublets in comparison with those found at RT.
- a higher percentage for $Fe_3O_4$ (m) than that assigned to the singlet of $Fe_3O_4$ (sp).

Therefore, we can conclude that at 298 K the doublets include: $Fe^{2+}$ in silica tetrahedral and octahedral sites and $Fe_3O_4$ (sp) with a crystal size lower than 6 nm.

In $c$-$Fe/SiO_2$(N) at 298 and 15 K, only a sextet is observed with hyperfine parameters characteristic of $\alpha$-$Fe^0$ (m).

The experimental $O_2$ uptake necessary for the complete re-oxidation of the reduced samples are shown in Table II. There is a good agreement between these values and the theoretical $O_2$ uptake calculated from the percentages of each species obtained from MS at 15 K assuming equal free recoilless factors for all iron species. The highly sensitive cross-checking of volumetric oxidation results with the MS of the catalysts is the only reliable method to verify the Mössbauer species assignments.

Table II shows the CO chemisorption values of the reduced samples obtained in a conventional static volumetric equipment. Using these values as described in [11], the average volumetric-superficial diameters ($d_{VA}$) of $Fe^0$ were estimated. In $c$-$Fe/SiO_2$(A) the $d_{VA}$ obtained was very small (2.2 nm). This result is in agreement with the $Fe^0$ (sp) signal detected by Mössbauer spectroscopy and with Bødker *et al.* [6] results mentioned above. It is important to note that the metallic crystal size obtained agrees with the size estimated for the oxide crystals in the precursor (lower than 3 nm). On the other hand, the $Fe^0$ $d_{VA}$ in $c$-$Fe/SiO_2$ (N) is about 10 times higher than in the sample calcined in air (29 nm). This value is coherent with that obtained for its precursor.

The important behavior difference (crystal size and reducibility) observed between both samples could be explained using the solid–solid wetting phenomenon. The wetting thermodynamical description can be applied to the solid–solid interface [12]. A schematic representation of wetting and spreading is

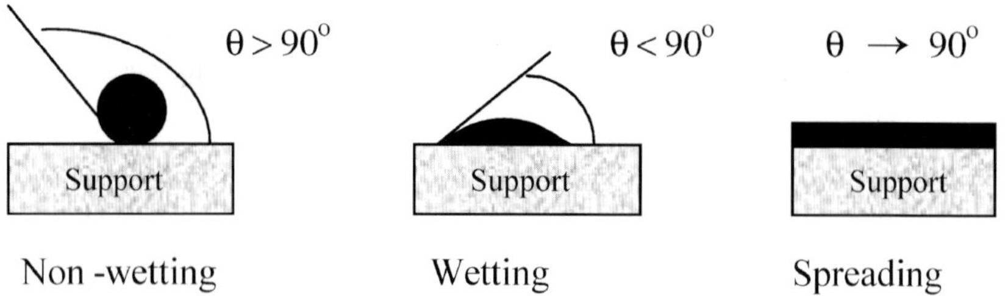


*Figure 2.* Scheme of the solid–solid wetting.

shown in Figure 2, where θ is the contact angle between the two phases that are contacting each other. θ is defined by the Young's equation:

$$\gamma_{ag} \cos\theta = \gamma_{sg} - \gamma_{as}$$

where $\gamma_{ij}$ denotes the interface free energy per unit area between: 'ag' (active phase-gas), 'sg' (support-gas) and 'as' (active phase-support).

Under equilibrium conditions:

θ > 90°, the active phase does not wet the support.
θ < 90°, wetting occurs.
θ → 0, spreading occurs.

The overall change in interfacial energy $\Delta F$ is given by:

$$\Delta F = \gamma_{ag}\Delta A_a - \gamma_{sg}\Delta A_s + \gamma_{as}\Delta A_s$$

Assuming equal $|\Delta A|$, being $\Delta A$ the changes in surface-interfaces areas:

$$(\gamma_{ag} + \gamma_{as})\cdot\langle\cdot\gamma_{sg}$$

Considering that the $\gamma_{as}$ is given by:

$$\gamma_{as} = \gamma_{ag} + \gamma_{sg} - U_{as}$$

where $U_{as}$ is the total interaction energy per unit interface area between the support and the active phase. Combining the two last equations:

$$U_{as}\rangle 2\gamma_{ag};$$

spreading occurs when this relation is satisfied.

In spite of the $\gamma_{ag}$ values are not always known, it has been established that the condition for spreading is more easily satisfied in oxidant atmosphere. Overbury *et al.* [13] found that $\gamma_{FeO-gas}$ increases 6.4 times when the $O_2$ concentration in the gas phase decreases from 0.25% to 0.001%. Therefore, in *p*-Fe/$SiO_2$(A) the spreading conditions could be satisfied leading to the production of islands like as 'raft' of amorphous iron oxide. On the other hand, it is possible that $\Delta F > 0$ in the absence of $O_2$ (*p*-Fe/$SiO_2$(N)), obtaining active phase particles on the support with sizes larger than the pore diameters in order to decrease the free energy of the system.

These important structural differences found between both precursors lead to the different reducibility degree and crystal metallic size in the reduced samples.

## 4. Conclusions

In the present work we have demonstrated that changing only one operative condition: $O_2$ content in the calcination atmosphere, important modifications in the structural properties of the reduced samples occur. In this way, it has been possible to obtain 100% of metallic iron with a particle diameter 10 times higher when the atmosphere changes from air to $N_2$. It is important to remark that several authors have reported the impossibility to obtain a complete reduction of iron species in $Fe/SiO_2$ systems with iron content lower than 10% (wt/wt) [14, 15]. However, the present results show that a 100% of reducibility is reached if a critical oxide diameter is overcome, even with 5% wt/wt of Fe content.

## Acknowledgements

The authors acknowledge support of this work by Consejo Nacional de Investigaciones Científicas y Técnicas (PIP 2853, PIP 2133), ANPCyT (PICT 14-11267), Comisión de Investigaciones Científicas de la Provincia de Buenos Aires and Universidad Nacional de La Plata, Argentina.

## References

1. Marchetti S. G., Cagnoli M. V., Alvarez A. M., Gallegos N. G., Bengoa J. F., Yeramián A. A. and Mercader R. C., *J. Phys. Chem. Solids* **58** (1997), 2119.
2. Bengoa J. F., Marchetti S. G., Cagnoli M. V., Alvarez A. M., Gallegos N. G. and Mercader R. C., *Hyperfine Interact. (C)* **5** (2003), 483.
3. Bengoa J. F., PhD Thesis Facultad de Ciencias Exactas, U. N. L. P. Cap. 2 (2003), 56.
4. Marchetti S. G. and Mercader R.C., *Hyperfine Interact.* **148–149** (2003), 275.
5. Clausen B. S., Topsøe H. and Mørup S., *Appl. Catal.* **48** (1989), 327.
6. Bødker F., Mørup S., Oxborrow C. A., Linderoth S., Madsen M. B. and Niemantsverdriet J. W., *J. Phys., Condens. Matter* **4** (1989), 6555.
7. van Lierop J. and Ryan D. H., *Phys. Rev., B* **63** (2001), 64406.
8. Berry F. J., Skinner S. and Thomas M. F., *J. Phys., Condens. Matter* **10** (1998), 215.
9. Mørup S., Topsoe H. and Lipka J., *J. Phys. Colloq.* **C6**(37) (1976), 287.
10. Lehlooh A.-F. and Mahmood S. H., *J. Magn. Magn. Mater.* **151** (1995), 163.
11. Cagnoli M. V., Marchetti S. G., Gallegos N. G., Alvarez A. M., Mercader R. C. and Yeramián A. A., *J. Catal.* **123** (1990), 21.
12. Knözinger H. and Taglauer E., In: Erte G., Knözinger H. and Weitkamp J. (eds.), *Handbook of Heterogeneuos Catalysis*, Wiley-VCH, Weinheim, 1997, p. 216.
13. Overbury S. H., Bertrand P. A. and Somorjai G. A., *Chem. Rev.* **75** (1975), 547.
14. Raupp G. B. and Delgass W. N., *J. Catal.* **58** (1979), 337.
15. Malathi N. and Puri S. P., *J. Phys. Soc. Jpn.* **31** (1971), 1418.

Hyperfine Interactions (2005) 161:11–19
DOI 10.1007/s10751-005-9187-z


# NMR and Mössbauer Study of $Al_2O_3$–$Eu_2O_3$

N. NAVA[1,*], P. SALAS[1], M. E. LLANOS[1], H. PÉREZ-PASTENES[2] and T. VIVEROS[2]
[1]*Instituto Mexicano del Petróleo, Programa de Ingeniería Molecular, Eje Central Lázaro Cárdenas No. 152, Col Atepehuacan, C.P. 07730, México D. F.; e-mail: tnava@imp.mx*
[2]*Departamento de IPH, Universidad Autónoma Metropolitana-Iztapalapa, Av. San Rafael Atlixco 186, Col. Vicentina, C.P. 09340, México D. F.*

**Abstract.** Alumina–europia mixed oxides with 5 and 10 wt.% $Eu_2O_3$ were studied by Mössbauer spectroscopy, $^{27}Al$ MAS-NMR and X-ray diffraction (XRD). The samples were prepared by the sol–gel technique. The XRD patterns for the calcined samples show a broad peak around 2θ = 30° which is assigned to the $Eu_2O_3$; after treatment with hydrogen at 1073 K no reduction to $Eu^{+2}$ or $Eu^0$ was observed. The NMR spectra show three peaks, which are assigned to the octahedral, pentahedral and tetrahedral aluminum sites; the intensity of each peak depends on the concentration of europium ions. The Mössbauer spectra of the calcined samples show a single peak near zero velocity which is attributed to the $Eu^{+3}$; after $H_2$ treatment at 1073 K similar spectra were obtained, suggesting $Eu^{+3}$ is not reducibly at this temperature.

**Key Words:** Mössbauer spectroscopy, nuclear magnetic resonance, X-ray diffraction.

## 1. Introduction

Rare earth oxides (REO) have been studied as catalytic promoters due to the generation of reactive sites, improvement of thermal stability and promoting the generation of basic sites [1, 2]. Alumina-based mixed oxides have been widely studied with different types of elements (Si, Zr, Ti, etc.) forming several solid solutions at certain concentrations; in the case of europium-promoted oxides prepared by impregnation method, $Eu_2O_3$ neutralizes acid sites on $Al_2O_3$ due to presence of hydroxyl groups and oxygen anion acting as base sites on the $Eu_2O_3$/$Al_2O_3$ supports [2]. These properties make appropriate these supports for oxidation reactions, hydrogenation–dehydrogenation of hydrocarbons, dehydration and dehydrogenation of alcohols, olefin isomerizations, paraffin cracking, and $NO_x$ elimination [3]. Furthermore, the preparation of catalytic mixed oxides by sol–gel chemistry provides a degree of control over the intimacy of molecular scale mixing that is not available by other techniques [4].

* Author for correspondence.

The $\gamma$-$Al_2O_3$ has many appealing properties that makes it interesting for applications in catalyst. The $\gamma$-$Al_2O_3$ structure is very closely related to that of Mg-spinel, which has the primitive unit cell $Mg_2Al_4O_8$, In Mg-spinel, the oxygen atoms are cubic close-packed, the Mg atoms are tetrahedrally coordinated by oxygen, and the Al atoms are octahedrally coordinated by oxygen.

Among the lanthanides europium is one of the most interesting elements to be studied, systematic studies of catalytic reactions on europium oxides are possible. $Eu_2O_3$ crystal possesses monoclinic and body centered cubic (bcc) structure.

In this work, we have prepared alumina–europia mixed oxides by the sol–gel technique and characterized by X-ray diffraction, nuclear magnetic resonance (NMR) and Mössbauer spectroscopy. The latter is a unique technique for investigating structural and dynamics aspects of solids on atomic scale [2, 5, 6].

## 2. Experimental

### 2.1. SAMPLES PREPARATION

$Al_2O_3$–$Eu_2O_3$ mixed oxides were prepared with 5 and 10 wt.% $Eu_2O_3$ as described: 1) aluminum tri-*sec*-butoxide was dissolved in excess of 2-propanol at 278 K, 2) $HNO_3$ solution was added slowly together with a solution of europium (III) nitrate hydrate in 2-propanol, 3) the gels obtained were aged 24 h, dried in vacuum at room temperature and calcined at 973 K, and 4) the samples were treated in 10% $H_2$/Ar (60 $cm^3$/min) at 1073 K for 2 h.

### 2.2. CHARACTERIZATION TECHNIQUES

#### 2.2.1. *XRD*

X-ray diffraction patterns were obtained with a Siemens D500 diffractometer equipped with a copper anode X-ray tube and a diffracted beam monochromator ($K\alpha$ = 1.5405 Å) and covering a range of diffraction angles $2\theta$ from 4° to 80°.

#### 2.2.2. *$^{27}$Al MAS-NMR*

The NMR spectra were recorded at 104.22 MHz, under MAS conditions in a 9.4 T static field with a Bruker-400 spectrometer. All measurements were carried out at 298 K; the spinning rate was 5 kHz. Chemical shift values are given relative to $AlCl_3$. Spectra were fitted with the WINFIT program [7].

#### 2.2.3. *Mössbauer spectroscopy*

Mössbauer spectra were collected at 4 K and at room temperature by using a constant acceleration spectrometer equipped with a $^{151}SmF_3$ source. The velocity

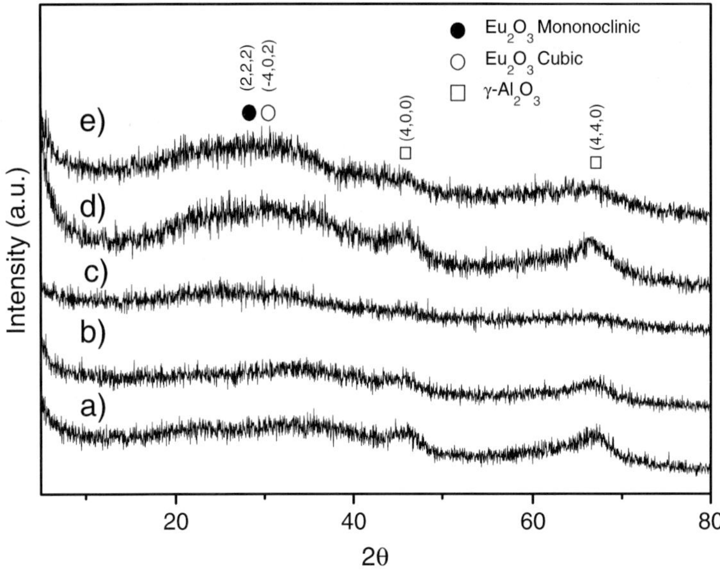


*Figure 1.* X-ray patterns of the samples: a) $\gamma$-$Al_2O_3$, b)5 wt.% $Eu_2O_3$–$Al_2O_3$, c) 10 wt.% $Eu_2O_3$–$Al_2O_3$, d) $H_2$-treated 5 wt.% $Eu_2O_3$–$Al_2O_3$, and e) $H_2$-treated 10 wt.% $Eu_2O_3$–$Al_2O_3$.

scale of the spectrometer was calibrated with the $^{57}$Fe Mössbauer resonance. Isomer shifts are reported relative to $EuF_3$. The samples were treated in flowing 10% $H_2$/Ar as described above and at 1073 K for 2 h before Mössbauer spectroscopy measurements. Reported isomer shifts are average values with uncertainties of 0.1 mm/s. The absorption spectra were fitted as a single line by using the NORMOS program [8].

## 3. Results

### 3.1. XRD

Figure 1 shows the XRD patterns of the $Al_2O_3$ and the $Al_2O_3$–$Eu_2O_3$ samples with 5 and 10 wt.% europium oxide, calcined and reduced. Figure 1a corresponds to the $\gamma$-$Al_2O_3$ pattern (JCPDS card 10-425, cubic structure; SG No 227). For the calcined samples containing 5 and 10 wt.% $Eu_2O_3$, the XRD signals of the $\gamma$-$Al_2O_3$ decrease, exhibiting only a weak pattern (Figure 1b,c), suggesting that europium oxide is well dispersed to the framework of the $\gamma$-$Al_2O_3$. When the samples were treated with $H_2$ (Figure 1d,e), a broad peak appeared around $2\theta = 30°$, that could be assigned to $Eu_2O_3$ according to JCPDS cards 12-384 (monoclinic structure; SG No. 12) and 12-393 (cubic structure; SG No. 206), which show main peaks at $2\theta = 28.51$ and 30.14 for $Eu_2O_3$. The presence of the peak around $2\theta = 30.14$ suggests a segregation and growth

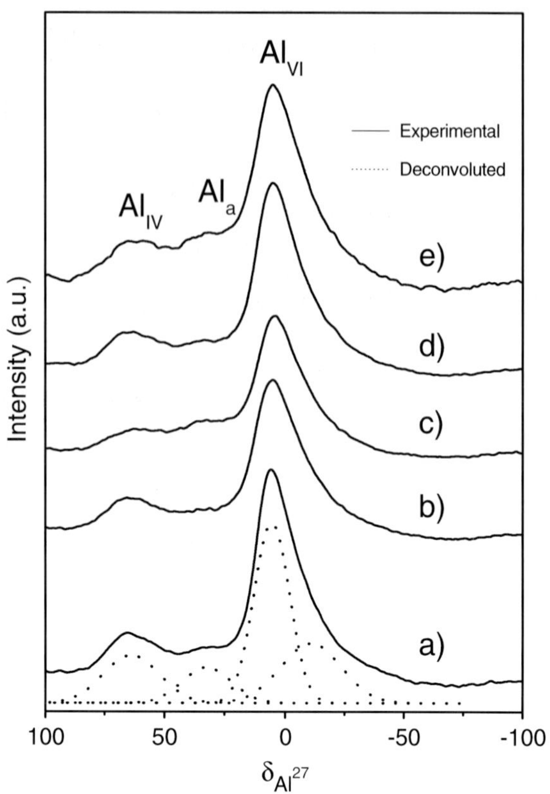


*Figure 2.* NMR spectra of the samples: a) $\gamma$-$Al_2O_3$, b) 5 wt.% $Eu_2O_3$–$Al_2O_3$, c) 10 wt.% $Eu_2O_3$–$Al_2O_3$, d) $H_2$-treated 5 wt.% $Eu_2O_3$–$Al_2O_3$, and e) $H_2$-treated 10 wt.% $Eu_2O_3$–$Al_2O_3$.

*Table I.* NMR parameters of $Al_2O_3$–$Eu_2O_3$ samples

| Sample $Eu_2O_3$ (wt.%) | Tetrahedral | | Pentahedral | | Octahedral | |
|---|---|---|---|---|---|---|
| | $\delta$ (ppm)[a] | Relative amounts (%)[b] | $\delta$ (ppm)[a] | Relative amounts (%)[b] | $\delta$ (ppm)[a] | Relative amounts (%)[b] |
| $\gamma$-$Al_2O_3$ | 63.99 | 20.20 | 31.89 | 9.48 | 6.09 | 70.32 |
| 5-Calcined | 64.21 | 19.27 | 33.09 | 9.51 | 5.85 | 71.22 |
| 5-Reduced | 63.97 | 16.97 | 32.77 | 10.93 | 5.38 | 72.10 |
| 10-Calcined | 62.54 | 12.76 | 32.01 | 19.37 | 4.42 | 67.87 |
| 10-Reduced | 63.17 | 16.37 | 33.49 | 13.09 | 5.70 | 70.54 |

[a] The estimated errors on $\delta$ are 1.0 ppm.

[b] Relative errors on integrated intensities are 4.0%.

process of $Eu_2O_3$ crystallites due to the hydrogen presence without europium reduction. As the concentration of $Eu_2O_3$ is increased, the crystallite size increases.

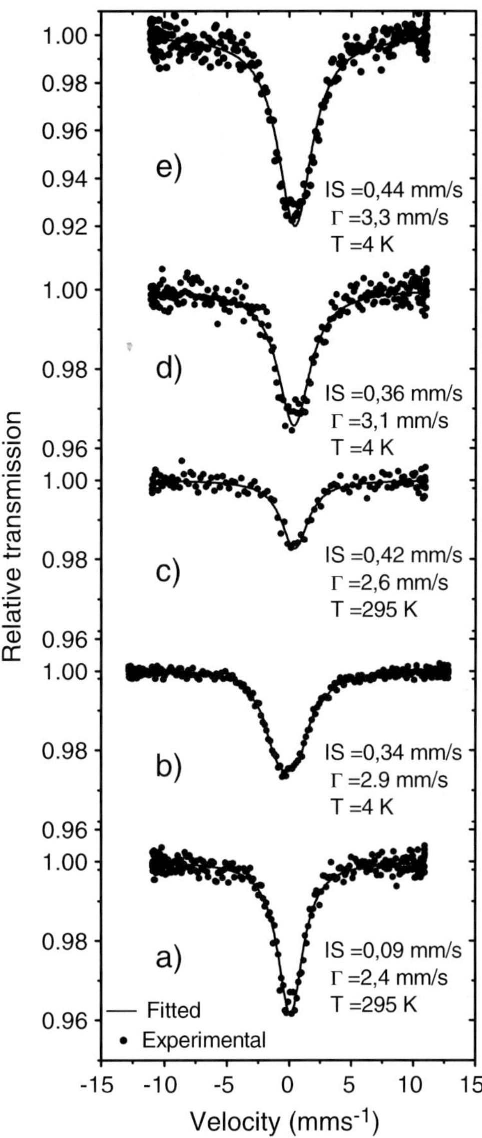


*Figure 3.* Eu-Mössbauer spectra of the samples: a) $Eu(NO_3)_3{\cdot}xH_2O$, b) 5 wt.% $Eu_2O_3$–$Al_2O_3$, c) 10 wt.% $Eu_2O_3$–$Al_2O_3$, d) $H_2$-treated 5 wt.% $Eu_2O_3$–$Al_2O_3$, and e) $H_2$-treated 10 wt.% $Eu_2O_3$–$Al_2O_3$.

## 3.2. NMR

The $^{27}$Al NMR spectra of γ-$Al_2O_3$, 5 and 10 wt.% $Eu_2O_3$ calcined and reduced samples are shown in Figure 2. The spectrum of the γ-$Al_2O_3$ shows three peaks, which are assigned to the octahedral, pentahedral and tetrahedral aluminum sites [9–12]. The NMR spectrum of the 5 wt.% $Eu_2O_3$–$Al_2O_3$ shows a decrease of the spectral area of the tetrahedral Al with respect to the γ-$Al_2O_3$, the pentahedral Al

*Table II.* Mössbauer parameters of $Al_2O_3$–$Eu_2O_3$ samples

| $Eu_2O_3$ (wt.%) | Pretreatment | IS (mm/s) | FWHM (mm/s) |
|---|---|---|---|
| 5 | Air | 0.34 | 2.9 |
| | Hydrogen | 0.36 | 3.1 |
| 10 | Air | 0.42 | 2.6 |
| | Hydrogen | 0.44 | 3.3 |
| $Eu(NO_3)_3{\cdot}xH_20$ | – | 0.09 | 2.4 |
| $Eu_2O_3$ | – | 0.22 | 2.3 |

increases and almost no change of the octahedral Al as is observed in Table I. A further decrease of the tetrahedral Al and increase of pentahedral Al is observed in $H_2$-treated sample. When the $Eu_2O_3$ is increased to 10 wt.%, a change in the Al spectral contribution is observed; a drastic decrease of the tetrahedral Al and an increase of the pentahedral Al are also observed. After treatment with $H_2$, an increase of tetrahedral Al and a decrease of pentahedral Al is observed. The relative amounts of each Al species were determined by deconvolution from the spectral area.

### 3.3. MÖSSBAUER SPECTROSCOPY

The Mössbauer spectra of the europium (III) nitrate hexahydrate, 5 and 10 wt.% $Eu_2O_3$ calcined and reduced samples are shown, in Figure 3 and the parameters involved can be seen in Table II. All the spectra show a broad absorption peak near zero velocity and as is well known it is difficult to measure the quadrupole splitting in $Eu^{+2}$ and $Eu^{+3}$ compounds because of the electronic configuration; $Eu^{+2}$ has $4f^7$ configuration with an electronic ground state of $^8S_{7/2}$, which cannot produce a valence-electron contribution to the electric field gradient; $Eu^{+3}$ has a $4f^6$ configuration and a ground state of $^7F_0$ which has zero total angular moment. So we analyze the peak as a single line [13]. The Mössbauer spectrum at room temperature of the europium (III) nitrate hexahydrate is shown in Figure 3a and as expected shows to be trivalent europium with an isomer shift of 0.09 mm/s and a line width of 2.4 mm/s. The Mössbauer spectra of both calcined samples (Figure 3b and c) exhibited a broad asymmetric peak, indicating that europium occupies multiple sites within the γ-$Al_2O_3$ framework, this is evident on the 5 wt.% $Eu_2O_3$ calcined sample, which shows a shoulder around 0.6 mm/s. In agreement to the IS, the central peak is attributed to a state oxidation of $Eu^{+3}$ (IS of $Eu_2O_3$ = 0.22 mm/s, FWHM = 2.30 mm/s) and the structure is assigned to the $Eu_2O_3$ phase. After treatment with $H_2$ at 1073 K, the samples show a similar spectrum as that calcined. As can be seen from Figure 4, no trace of $Eu^{+2}$ and $Eu^0$ were obtained, the spectrum taken at 40 mm/s and $T = 4$ K corresponds to the 10 wt.% $Eu_2O_3$ reduced sample.

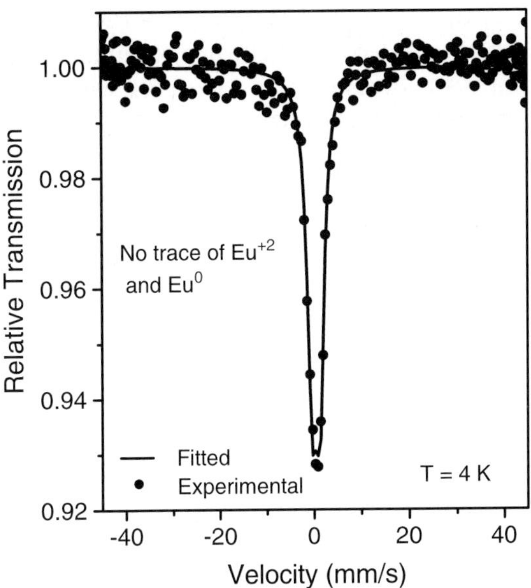


*Figure 4.* Mössbauer spectrum of the $H_2$-treated 10 wt.% $Eu_2O_3$–$Al_2O_3$ sample, taken at $T = 4$ K and $V = 40$ mm/s.

The difference in intensity between the 10 wt.% $Eu_2O_3$ calcined and reduced is due to the temperature at which they were taken; the calcined sample was measured at room temperature while the reduced sample was measured at 4 K.

## 4. Discussion

After calcination, europia is dispersed on the $\gamma$-$Al_2O_3$, as is observed by the broadening of the XRD patterns in the range $2\theta = 10°$ to $35°$; however for the sample containing 10 wt.% $Eu_2O_3$ exhibited the weak XRD patterns, suggesting that $Eu^{+3}$ ions occupy more sites on the $\gamma$-$Al_2O_3$ support than the sample containing 5 wt.% $Eu_2O_3$. After reduction treatment the intensity of the peaks in the region $2\theta = 30°$ is more intense than the calcined samples; it suggests that after reduction with hydrogen the $Eu_2O_3$ crystal size increases, and the fact that the XRD patterns for samples with 5 and 10 wt.% $Eu_2O_3$ show a different pattern suggests that the europium ions are segregated as small $Eu_2O_3$ crystalline particles in strong interaction with the $\gamma$-$Al_2O_3$ framework.

The extensive literature on $\gamma$-$Al_2O_3$ has established that the spinel structure has a proportion of octahedral Al sites to tetrahedral Al sites of 66%–33%. The NMR studies have revealed another type of Al, which is denominated pentahedral Al and is attributed to aluminum atoms at the surface of the particles [11]. In these samples, NMR results indicate there is a significant amount of pentahedral Al that modifies the proportion of tetrahedral–octahedral sites. The NMR results revealed also that when europium is added an interchanged between tetrahedral Al and octahedral Al occurs while octahedral Al remains almost constant. This type of structure is referred as a defective spinel in which vacancies are located at tetrahedral sites.

For the octahedral peak two subspectra were used to fit the spectrum due to the asymmetry of the peak.

After $H_2$ treatment at 1073 K similar Mössbauer spectra are obtained, suggesting $Eu^{+3}$ is not reducible at this temperature. It has been observed that $Eu_2O_3$ is stable towards $H_2$ reduction below 1473 K, however in some works it has been reported [2] it can be reduced to $Eu^{+2}$ in $H_2$ at 673–773 K when supported on $\gamma$-$Al_2O_3$. However, according to our Mössbauer spectra at 40 mm/s, no evidence of $Eu^{+2}$ or $Eu^0$ was obtained.

$Eu^{+3}$ is the most common state of oxidation in ionic salts. This fact is manifested in the europium nitrate hexahydrated salt that in according to the IS the state of oxidation of europium ion is +3. As it was pointed before the $Eu^{+3}$ ion has a valence shell consisting of six electrons in 4f orbital; an increase of the IS indicates an increase in the electron density at the nucleus of $^{151}Eu$ with an increase in the number of 6s electrons [14]. For the calcined and reduced samples an increase in the IS is observed as the concentration of $Eu_2O_3$ is increased, this is caused by an increase in the covalence of $Eu^{+3}$ ions with the $Al^{+3}$ ions.

This result is consistent with the XRD patterns in which only an increase of the intensity of the peak at $2\theta = 30°$ is observed after reduction with $H_2$. It means that $Eu_2O_3$ phase increases the crystal size but is in strong interaction with the $\gamma$-alumina, which prevents europium reduction. Furthermore the FWHM of the calcined and $H_2$-treated samples increases with respect the europium salt, indicating an increase in the asymmetry, due to the influence of the $\gamma$-$Al_2O_3$.

## 5. Conclusions

The XRD patterns indicate that europium ions are present as small $Eu_2O_3$ particles dispersed on the $\gamma$-$Al_2O_3$.

The Mössbauer results indicate that europium ions are present as the trivalent oxidation state, and after reduction no evidence of $Eu^{+2}$ or $Eu^0$ was observed. After calcination and $H_2$ treatment, an increase in the Isomer Shift is observed; this indicates an increase in the covalence of $Eu^{+3}$ ions with the $\gamma$-alumina.

The NMR spectra show the presence of three aluminum sites, tetrahedral, pentahedral and octahedral for the $\gamma$-$Al_2O_3$; after calcination and $H_2$ treatment, the amount of tetrahedral Al decreases while the pentahedral Al increases; hence europium ions modify mainly tetrahedral sites of the $\gamma$-$Al_2O_3$.

## References

1. Zhang G., Hattor H. and Tanabe K., *Appl. Catal.* **36** (1988), 189.
2. Shen J., Lochhead M. K., Chen Y. and Dumesic J. A., *J. Phys. Chem.* **99** (1995), 2384.
3. Rosynek M., *Catal. Rev., Sci. Eng.* **16** (1977), 111.
4. Miller J. and Ko E., *Catal. Today* **35**(3) (1977), 269.
5. Philip N. and Delgass W., *J. Catal.* **33** (1974), 219.
6. Boolchand, Mishra P. K., Raukas M., Ellens A. and Schmidt P., *Phys. Rev., B* **65** (2002), 134429.
7. Massiot D., WINFIT Programme, Bruker-Franzen Analytic Gmbh, Berlin, 1993.
8. http://www.wissel-instruments.de/produkte/software.html.
9. Fripiat J., Alvarez L., Sánchez J., Martínez E., Saniger J. and Sánchez N., *Appl. Catal., A Gen.* **215** (2001), 91.
10. Blumenfeld A. and Fripiat J., *J. Phys. Chem., B* **101** (1977), 6670.
11. Pecharromán C., Sobrados I., Iglesias J., Gonzalez-Carreño T. and Sanz J., *J. Phys. Chem., B* **103** (1999), 6160.
12. Wang J., Bokhimi X., Morales A., Novaro O., López T. and Gómez R., *J. Phys. Chem., B* **103** (1999), 299.
13. Greenwood N. and Gibb T., *Mössbauer Spectroscopy*, Chapman Hall Ltd, London, 1971.
14. Fujita K., Tanaka K., Hirao K. and Soga N., *J. Am. Ceram. Soc.* **81** (7) (1998), 1845.

Hyperfine Interactions (2005) 161:21–31
DOI 10.1007/s10751-005-9188-y


# From Ore to Tool – Iron Age Iron Smelting in the Largest and Oldest Meteorite Crater in the World

F. B. WAANDERS[1,*], L. R. TIEDT[2], M. C. BRINK[2] and A. A. BISSCHOFF[2,#]
[1]*School of Chemical and Minerals Engineering, North-West University, Potchefsroom, 2520 South Africa; e-mail: chifbw@puk.ac.za*
[2]*Laboratory for Electron Microscopy, North-West University, Potchefstroom, 2520, South Africa*

**Abstract.** The Vredefort Impact Structure in South Africa is the biggest and oldest remnant meteorite impact crater in the world where various ancient cultures thrived. In this paper some light will be shed on the Iron Age, iron smelting aspects of the people that inhabited the area and the results of a laboratory study of iron artefacts and a possible source of iron ore in the region is given. A sectional piece from a hoe manufactured in a small bloomery furnace was polished and etched and subsequently analyses with SEM and Mössbauer techniques were obtained. The hoe has a typical cast iron composition (2.9% C, 0.1% Mn, 0.4% Si, 0.4% P and 96.2% Fe, all wt.%) and contains many slag inclusions with wustite dendrites. The Mössbauer spectrum consisted of iron (86%), wustite (5%) and oxihydroxide (9%) and the thin (200 μm) corrosion layer consisted of hematite (55%) and oxihydroxides (45%). At a furnace site, various slag clumps (26.3% C, 24.8% $SiO_2$, 11.3% $Al_2O_3$, 1.3% $P_2O_5$, 1.0% $K_2O$, 0.4% CaO and 30.2 FeO, all wt.%, average of four samples) and iron nodules (7.6% C, 6.0% Mn, 4.3% Si, 1.4% Al, 80.7% Fe, all wt.%) were found. The Mössbauer spectrum of the slag consisted of iron (7%), magnetite (56%), fayalite (2%) and oxihydroxides (35%) and that of the iron nodules yielded iron (28%), wustite (12%), magnetite (20%) and oxihydroxides (40%). A possible ore source containing 84% FeO, 7% of $Al_2O_3$ and $SiO_2$ (all in wt.%) and minor impurities is located a few kilometers from the furnace site, yielding a Mössbauer spectrum consisting of hematite (70%) and oxihydroxides (30%).

**Key Words:** $^{57}$Fe-Mössbauer iron ore, cast iron, iron smelting, SEM analyses, slag, spectroscopy.

## 1. Introduction

South Africa and Mexico have various things in common – famous meteorite craters and archeological sites, just to mention two. The Chicxulub giant crater [1] of about 180 km across is buried under the younger sediments of the Yucatán Peninsula and was most probably the catastrophic meteorite impact that led to the mass extinction of the dinosaurs. Not related to the impact crater structure, are the numerous Mayan archeological sites in the region, dating back hundreds of years. The Vredefort Impact Structure in South Africa, of more than 300 km in

* Author for correspondence.
# Retired.

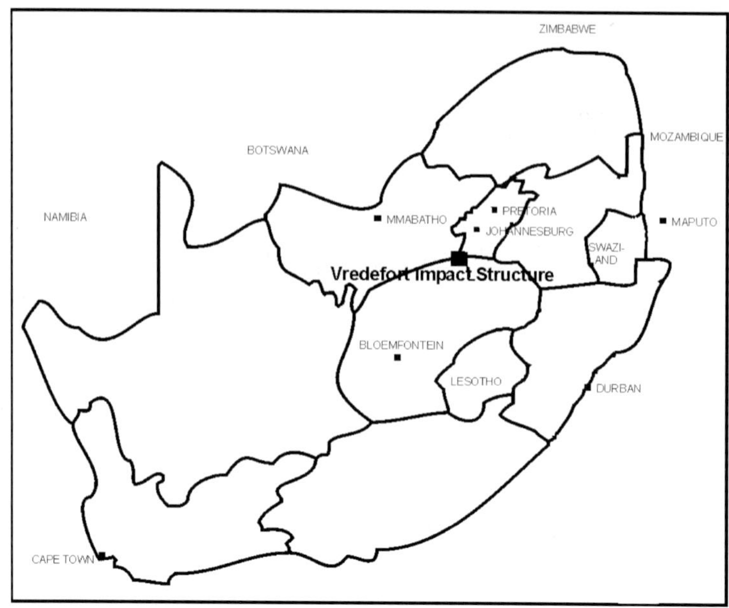


*Figure 1.* Locality map of the Vredefort Impact Structure found in South Africa.

diameter, however, is the biggest and oldest remnant meteorite impact crater in the world [2] where various ancient cultures thrived [3]. The time of the impact has been dated back to 2GA ago, which is thus much older than the Chicxulub impact of 65 million years ago.

In the 1930s, the inner circular portion of the Vredefort crater was included into a world wide list of structures that possibly owed their formation to meteoritic impacts and in 2002 application was made to proclaim the Vredefort Impact Structure as a World Heritage Site. The locality of the Vredefort Impact crater is shown in Figure 1 and a simplified geological map is shown in Figure 2, which shows the effects of the impact at 200 km from ground zero, beyond the town of Lichtenburg, South Africa, to the north west of the centre of the Structure. The development of the structure in the first few seconds after impact can be deciphered from the deformation of the rock masses that were displaced. The thick layers of mainly sedimentary rock, some high in iron oxide content, which could be possible sources for iron ore, were broken during impact. These contain recognisable marker beds, which can now be mapped, and a reconstruction can be made of the way large rock blocks moved relatively to each other during and immediate after impact.

Contained within the more or less 15 km thick sandwich-like rock pile, are the Witwatersrand rocks, which contain the richest single gold deposit on earth. The various crustal movements during the few seconds of torture after impact are now reasonably well known and the complete sequence of the evolution of the structure can be found in [2]. Of importance is the fact that without the impact episode, ore deposits would not have been uplifted to the present surface level.

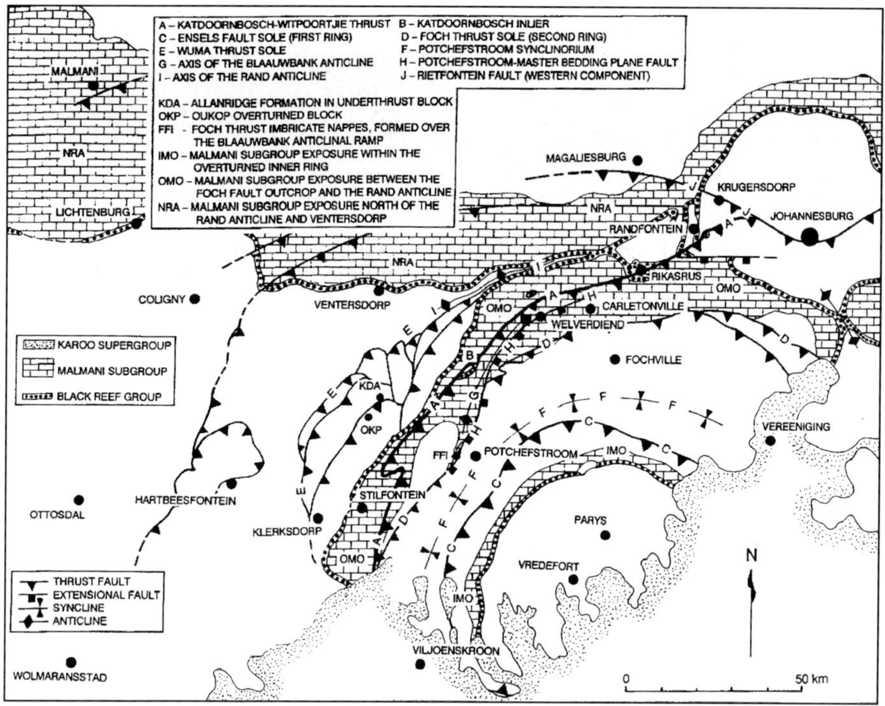


*Figure 2.* Simplified geological map of the Vredefort Impact Structure [4].

In this paper some light will be shed on the Iron Age iron smelting aspects of the people that inhabited the Vredefort area. The main evidence for metal workings are furnace remains, tuyères, crucible shreds, slag and of course the metal artefacts themselves. Based on historical and linguistic facts, Hromník [5] showed that nowhere is the connection of Indians mainly from southern India with early African reality felt as strongly as in the area of metallurgy. The people of India have a world-wide reputation as the most skilful iron-workers as they understood best the technique of preparing the right mixture of iron ore, carbon, air and other ingredients for successful fusion of a suitable iron for various purposes. From history it is known that the Indians lived amongst people form Africa from the 10th century AD [5]. In the mid-seventies radiocarbon dating of Iron Age sites gave the date of an iron bloomery furnace (production of iron in solid form at temperatures less than 1300°C as a result of the reduction of iron ore) from Melville Koppies, South Africa as 1060 ± 50 AD [5], thus of the same period when Indians lived in the region. Even older Iron Age finds of 300 AD were identified [6], indicating that the Africans could have learned the iron-smelting techniques from the Indians over many years. Furthermore, the iron artefacts from various South African Iron Age sites and periods show few differences and one can conclude that the production methods for traditional bloomery iron did not change much over the last 1000 years [7].

*Figure 3.* On the right hand side a poorly exposed furnace, showing only a small portion of the furnace visible, just above the hammer and on the right, typical agricultural implements manufactured from the iron produced in such a furnace.

The ancient agricultural tools found in South Africa fall mainly into two categories, namely those produced before the Europeans entered Africa and those made after trading with the European settlers some 500 years ago. Various investigations on settlements, copper and iron artefacts have been made and more than 200 papers, dating back to the end of the 19th century to the current, were published [8], but none focused on finds in the Vredefort area. The discovery of iron furnaces and metal artefacts in the area of the Vredefort Impact Structure indicates that the tools (Figure 3, right) investigated in this project, were manufactured in small bloomery furnaces of about 1 m in diameter and height (Figure 3, left), dating to a time before Europeans entered Africa.

## 2. Experimental

A hoe, found in the vicinity of the furnace, was presented by a local farmer for investigation. The stem was sectioned and polished according to standard metallographic methods. The sample was etched with nital and examined under the SEM and with Mössbauer spectroscopy. At the furnace site various slag and iron nodule samples were found and analysed with the same techniques. To try and find a possible source of iron ore it was assumed that the ore would not have been carried over vast distances and at least four possible iron-containing rocks from the area were collected and further investigated.

The Mössbauer spectra were obtained with the aid of a Halder Mössbauer spectrometer, capable of operating in conventional constant acceleration mode using either a proportional counter filled with Xe-gas to 2 atm or a backscatter-type gas-flow detector. A 50 mCi $^{57}$Co source, plated into a Rh-foil, was used to produce the $\gamma$-rays. The samples were analysed at room temperature and data collected in a multi-channel analyser (MCA) to obtain a spectrum of count rate against source velocity. A least-squares fitting program was used and by superimposing Lorentzian line shapes, the isomer shifts, quadrupole splitting

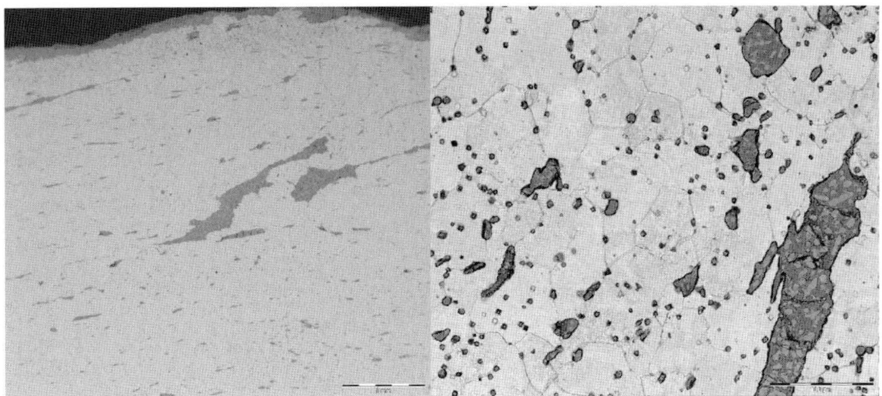

*Figure 4.* On the left the SEM microphotograph of the cross section of the hoe, showing the dark grey slag inclusions, more or less parallel to the outer surface and on the right an enlarged microphotograph showing the individual light ferrite grains, minor grain boundary cementite and the dark grey slag inclusions.

and/or hyperfine magnetic field of each constituent was determined with reference to the centroid of the spectrum of a standard α-iron foil at room temperature. The amount of each constituent present in a sample was determined from the areas under the relevant peaks.

A FEI Quanta 200 ESEM Scanning Electron Microscope fitted with an Oxford Inca 400 energy dispersive X-ray spectrometer (EDS) was used for the SEM analyses of the samples received.

## 3. Results and discussion

From the sectioned and polished part of the hoe the SEM analyses yielded the composition of the hoe as 2.9% C, 0.1% Mn, 0.4% Si, 0.4% P and 96.2% Fe, all wt.%, which is typical for a cast iron composition. The thin (200 μm) corrosion layer (see Figure 4) is clearly visible on the top surface and furthermore many slag inclusions can be seen. Equitant recrystallized ferrite grains, with grain sizes varying between ASTM 5 and ASTM 8 (the lager number corresponds to finer grain sizes) are associated with grain boundary cementite and pearlite. A statistical evaluation of the grain sizes from the microphotograph yielded an average grain size of 51 μm, or ASTM 6, (see Figure 4) similar to the results shown in [9]. The Vickers Hardness of the sample was found to be 70–100 HV, depending on the position where the measurement was taken. According to [7] the hardness of various samples tested, varied between 124 and 258 HV, thus a factor two higher than that found in the present investigation.

The slag inclusions are present in elongated stringers, arranged parallel to the direction of extension in hot working when the hoe shape was forged. Within these dark grey slag inclusions wustite dendrites are observed (see Figure 5)

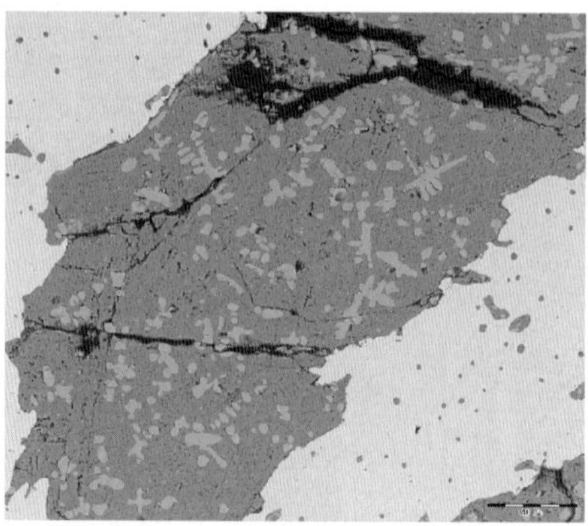

*Figure 5.* Dark grey, large slag inclusion with lighter grey wustite dendrites clearly visible with the very light grey corresponding to the ferrite and the voids are black.

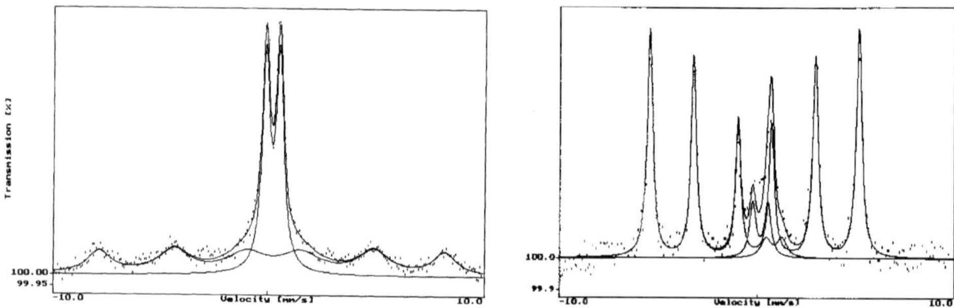

*Figure 6.* On the left the Mössbauer spectrum of the corrosion layer consisting of hematite and an oxihydroxide and on the right the spectrum of the hoe showing the iron sextet, the wustite and oxihydroxide doublets.

containing 72 wt.% Fe and 25 wt.% O from which an estimation composition of $Fe_{0.93}O$ for the wustite can be deduced. Furthermore, to have a wustite with 25 wt.% O, the temperature in the bloomery furnace must have been about 1100°C, as was found in reconstructed South African Iron Age furnace experiments [10].

The Mössbauer spectrum obtained for the hoe consisted of iron, wustite and an oxihydroxide, whilst the thin (200 μm) corrosion layer consisted of hematite and oxihydroxide (see Figure 6). The hematite and oxihydroxide are indicative of the corrosion that took place over the years. In [11] an estimation of the age of the tools investigated was made from the thickness of the corrosion layer and for a layer of 200 μm thick an estimated minimum weathering time of 22 years was calculated. In the present investigation however the tools and nodules were buried most probably under a layer of soil, at least 15 cm thick, which would have prevented exposure to the atmosphere and thus the corrosion rates, might have been totally different. The spectra are shown in Figure 6 and the Mössbauer parameters, which are consistent with literature [12], are tabulated in Table I.

*Table I.* Hyperfine interaction Mössbauer parameters of Fe-components found in this investigation

| Sample | Component | IS $mms^{-1}$ (±0.02) | QS $mms^{-1}$ (±0.02) | H Tesla (±0.3) | Relative intensity (%) |
|---|---|---|---|---|---|
| Hoe | Iron | 0.02 | 0.00 | 33.2 | 86 ± 2 |
| | Wustite | 0.95 | 0.81 | – | 5 ± 1 |
| | Oxihydroxide | 0.25 | 0.76 | – | 9 ± 1 |
| Hoe rust layer | Hematite | 0.34 | −0.09 | 50.0 | 55 ± 2 |
| | Oxihydroxide | 0.34 | 0.66 | – | 45 ± 2 |
| Iron nodule | Iron | −0.03 | −0.05 | 32.8 | 28 ± 2 |
| | Wustite | 0.88 | 0.85 | – | 12 ± 2 |
| | Magnetite | 0.41 | −0.12 | 49.1 | 20 ± 2 |
| | Oxihydroxide | 0.36 | 0.66 | – | 40 ± 2 |
| Slag nodule | Iron | 0.00 | −0.02 | 33.3 | 7 ± 2 |
| | Magnetite | 0.27 | 0.01 | 49.9 | 34 ± 2 |
| | | 0.58 | 0.07 | 46.2 | 22 ± 2 |
| | Fayalite | 1.13 | 2.77 | – | 2 ± 1 |
| | Oxihydroxide | 0.36 | 0.70 | – | 35 ± 2 |
| Diamictite | Hematite | 0.34 | −0.17 | 50.6 | 70 ± 2 |
| | Oxihydroxide | 0.33 | 0.57 | – | 30 ± 2 |
| Water Tower slate | Hematite | 0.35 | −0.14 | 51.9 | 42 ± 2 |
| | Magnetite | 0.31 | 0.01 | 49.5 | 14 ± 2 |
| | | 0.65 | −0.02 | 46.2 | 12 ± 1 |
| | Grunerite | 1.21 | 2.93 | – | 26 ± 2 |
| | | 1.11 | 1.56 | – | 6 ± 1 |
| Contorted bed | Hematite | 0.32 | −0.09 | 52.0 | 43 ± 2 |
| | Magnetite | 0.21 | −0.08 | 49.8 | 14 ± 2 |
| | | 0.73 | −0.02 | 46.7 | 11 ± 2 |
| | Grunerite | 1.21 | 2.95 | – | 26 ± 2 |
| | | 1.10 | 1.55 | – | 6 ± 1 |
| Specularite | Hematite | 0.35 | −0.18 | 50.7 | 67 ± 2 |
| | Oxihydroxide | 0.35 | 0.58 | – | 33 ± 2 |
| Ore lumps found at furnace | Hematite | 0.35 | −0.10 | 51.7 | 69 ± 2 |
| | Magnetite | 0.36 | 0.02 | 49.3 | 17 ± 1 |
| | | 0.62 | 0.07 | 46.1 | 8 ± 1 |
| | Oxihydroxide | 0.27 | 0.69 | – | 6 ± 1 |

Note: IS = Isomer shift relative to α-Fe, QS = Quadrupole splitting and H = Hyperfine magnetic field strength.

At the furnace site, various slag nodules (26.3% C, 24.8% $SiO_2$, 11.3% $Al_2O_3$, 1.3% $P_2O_5$, 1.0% $K_2O$, 0.4% CaO and 30.2 FeO, all wt.%, average of four samples) were found. The amount of iron present in the slag is within the range of 30%–57% $Fe^{2+}$or 10%–34% $Fe^{3+}$, as reported for various slags found in Southern Africa [13] and the present slag nodules can thus be recognized as a typical slag from Iron Age smelting furnaces. Furthermore the overall composition of the slags falls in the ranges given in [14]. In addition to the slag, iron nodules (7.6% C, 6.0% Mn, 4.3% Si, 1.4% Al, 80.7% Fe, all wt.%) were also found at the furnace site. Both the slag and iron nodules were buried under the

*Figure 7.* On the left two slag nodules and on the right two iron nodules of which the one nodule was ground to show the high metallic content of the nodule. The size of the nodules varies between 2 and 4 cm in diameter.

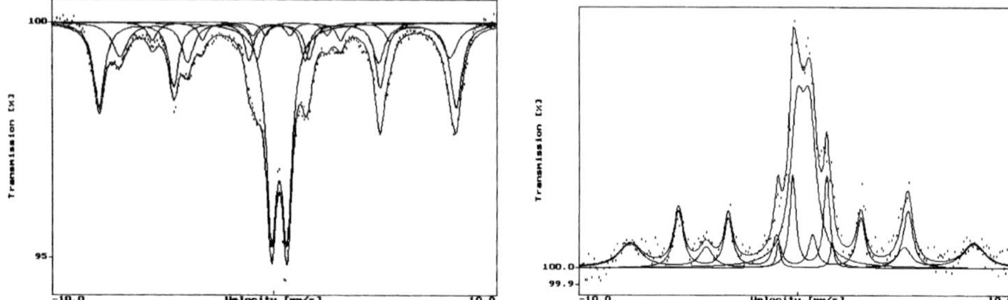


*Figure 8.* On the left the Mössbauer spectrum of the slag showing the two magnetite sextets, the iron sextet and the fayalite and oxihydroxide doublets. On the left the spectrum of the iron nodule containing iron, magnetite, wustite and oxihydroxide.

soil surface and were dug up for investigation. A clear difference in morphology between slag and iron nodules was observed, as can be seen from Figure 7. The most left slag nodule in Figure 7 clearly shows a flow pattern and was subjected to Mössbauer spectroscopy, yielding the spectra shown in Figure 8. A small amount of fayalite ($Fe_2SiO_4$) was present in the Mössbauer spectrum, in addition to the magnetite, iron and oxihydroxide, but the fayalite could not be detected in slag samples that did not show this flow pattern. The iron nodules were also subjected to Mössbauer spectroscopy and all the Mössbauer parameters of slag and iron nodules are shown in Table I and compare favorably with those found in literature [12].

From the Mössbauer spectrum of the iron nodule it is clear that the total amount of metallic iron is less than that found in the hoe but high enough to ensure the metallic nature of the nodule. The presence of the magnetite indicates that no complete reduction of this particular nodule has taken place or that, in addition to the oxihydroxide, it is indicative of the corrosion that has taken place. Due to unsatisfactory statistics and the relative low abundance of the magnetite, no attempt was made to resolve the broad peaks and only one sextet was fitted to the experimental data, leading to the best $\chi^2$ value to be obtained during fitting.

*Table II.* Chemical composition of the various iron ores found in the vicinity of the furnace

| Constituent % | Diamictite | Water tower slate | Contorted bed | Specularite | Ore at furnace |
|---|---|---|---|---|---|
| Silicon ($SiO_2$) | 7.45 | 49.39 | 48.81 | 11.51 | 32.41 |
| Aluminium ($Al_2O_3$) | 7.42 | 7.97 | 3.17 | 1.12 | 2.44 |
| Iron (FeO) | 84.23 | 40.51 | 40.75 | 85.63 | 63.68 |
| Magnesium (MgO) | – | 0.41 | 1.40 | – | 0.44 |
| Calcium (CaO) | – | 0.38 | 1.83 | 0.34 | 0.30 |
| Potassium ($K_2O$) | – | 0.26 | – | – | 0.29 |
| Manganese (MnO) | 0.28 | 1.08 | 4.04 | – | 0.44 |
| Phosphorus ($P_2O_5$) | 0.62 | – | – | – | – |
| Other | – | Zn (0.41) | – | S (1.39) | – |

In an attempt to find a possible source for the iron ore that could have been used to produce the iron artefacts, the iron bearing or ferruginous stratigraphic horizons in the vicinity of the furnace were investigated and mainly two prominent bands were identified. The one band being the so-called Water Tower slates and the other band the Contorted Bed. Both these horizons belong to the West Rand Group of sediments that form part of the gold bearing Witwatersrand rocks [15], which were uplifted to surface after the meteorite impact. The Water Tower slate is a band of between 3 and 5 m thick, heavy ferruginous with platy specularite ($Fe_2O_3$) present on cleavage planes. A diamictite in the vicinity of these slates, also belonging to the West Rand Group of sediments was also sampled due to the fact that it is highly ferruginous and can be described as a pebbly mudstone containing high amounts of iron. The Contorted Bed is a banded ironstone and for most of the distance over which it outcrops, an intrusive sill occurs in the middle of the bed. At another furnace site, a few kilometers from the furnace investigated, various ore lumps were left behind at the furnace site and were also sampled, but they are of unknown origin.

The chemical analysis of the ores, as determined by means of SEM analyses, is shown in Table II. Both the Water Tower slate and Contorted Bed have equal amounts of iron, but only half the amount found in the diamictite, specularite and two thirds compared to the ore lumps. The amount of iron present can however be sufficient to be used as an iron ore, but from the Mössbauer spectra (see Figure 9) it became clear that more than a third of the iron in the two samples is bound with the silica to form grunerite ($Fe_7Si_8O_{22}(OH_2)$), which would be difficult to reduce under archaeological bloomery conditions. Cummingtonite ($(Mg,Fe)_7Si_8O_{22}(OH_2)$) has similar Mössbauer parameters as grunerite, but the amount of Fe > Mg in the samples investigated and they should therefore be regarded as containing grunerite. Within the realm of the metamorphic aureole that formed due to the intrusion of an alkali granite in the area of the possible iron ores, the shales have been altered to amphibole-magnetite bearing rocks [15]. Since the mudstone contains very little silica and alumina, it was not metamorphosed in the

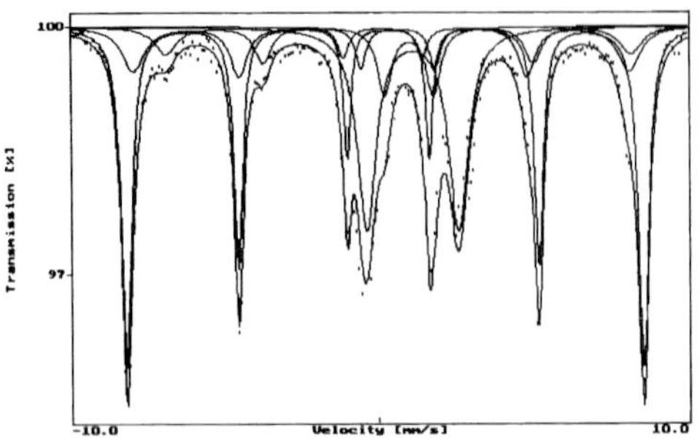


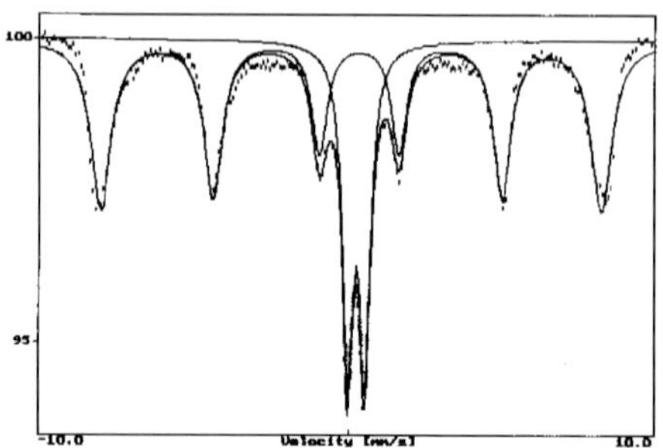


*Figure 9.* On the left the Mössbauer spectrum of the contorted bed, showing the hematite, magnetite and the two grunerite doublets, whilst on the right hand side the spectrum of the diamictite is shown, indicating the presence of hematite and oxihydroxide.

same way as the Water Tower slate and Contorted Bed and it could have been the source of iron as it consists mainly of hematite, suitable for the iron furnace process. The amount of specularite in the Water Tower slate is also limited to thin shiny plates on cleavage planes and would not provide sufficient ore for the production of many artefacts.

Finally the ore lumps were analyzed and from SEM analyses it had a similar composition as the diamictite. The Mössbauer spectrum was also almost identical to that of the diamictite, but from a geological point of view not from the same site, which needs further investigation.

## 4. Conclusion

This research project evolved from an initial geological interest in the Vredefort Impact Structure. Much more has to be done to unravel the mysteries surrounding the archaeological aspects of the area as only very little and sporadic detailed research so far has been done in the region. Hopefully the area will be declared a World Heritage Site soon, which might change this.

From the present investigation however, it can be stated that the hoe was manufactured during the Iron Age period, due to physical metallurgical observations such as the presence of the many slag inclusions, wustite dendrites, iron carbides on grain boundaries and the cold working textures present. The wustite was also observed via Mössbauer spectroscopy, but unfortunately no iron carbides were observed due to overlapping of the peaks in the Mössbauer spectra. Isotope techniques could elucidate more, but these techniques were unfortunately not available at the time of the investigation.

## Acknowledgements

Mr. Johannes van der Merwe and all the farmers who are most cooperative en helpful in our ventures in the Vredefort Impact Structure area are highly appreciated. Financial support from the North-West University and the National Research Foundation is appreciated.

## References


1. French B. M., In: *Traces of Catastrophe: A Handbook of Shock-Metamorphic Effects in Terrestrial Meteorite Impact Structures*. LPI Contribution No. 954, Lunar and Planetary Institute, Houston, 1998, pp. 120.
2. Brink M. C., Waanders F. B. and Bisschoff A. A., *Tectonophysics* **270** (1997), 83–114.
3. Maggs T. M. O'C., *Iron Age Communities of the Southern Highveld*, Council of the Natal Museum, Pietermaritzburg, South Africa, 1976, pp. 326.
4. Brink M. C., Waanders F. B. and Bisschoff A. A., *S. Afr. J. Geol.* **103**(1) (2000), 15–31.
5. Hromník C. R., *Indo-Africa Towards a New Understanding of the History of Sub-Saharan Africa*, Juta and Co Ltd, P.O. Box 123, Kenwyn, 7790, South Africa, 1981, p. 168.
6. Mason R. J., *SAIMM Journal* (Jan. 1974), 211–216.
7. Friede H. M. and Steel R. H., *S. Afr. Archaeol. Bull.* **41** (1986), 81–86.
8. Miller D. E. and Maggs T. M. O'C., *Pre-Colonial Metalworking in Africa, Especially Southern Africa: A Bibliography: 1–67*. Department of Archaeology, University of Cape Town (African Studies Library), 1994, p. 100.
9. Miller D. and Whitelaw G., *S. Afr. Archaeol. Bull.* **49** (1994), 79–89.
10. Friede H. M., Hejja A. A., Koursaris A. and Steel R. H., *J.S. Afr. Inst. Min. Metall.* **84**(9) (Sept. 1984), 285–297.
11. Gordon R. B. and van der Merwe N. J., *Archaeometry* **26**(1) (1984), 108–127.
12. Stevens J. G., Khasanov A. M., Miller J. W., Pollak H. and Li Z. (eds.), *Mössbauer Mineral Handbook*, University of North Carolina, Asheville, USA, 1998, p. 527.
13. Friede H. M., *J.S. Afr. Inst. Min. Metall.* (June 1977), 224–232.
14. Friede H. M., Hejja A. A. and Koursaris A., *J.S. Afr. Inst. Min. Metall.* (Feb. 1982), 38–52.
15. Bisschoff A. A., In: van Vuuren C. J., Moen H. F. G. and Minnaar C. L. J. (eds.), *The Geology of the Vredefort Dome, Explanation of Geological Sheets*, Council for Geoscience, Pretoria, South Africa, 1999, p. 49.

Hyperfine Interactions (2005) 161:33–42
DOI 10.1007/s10751-005-9189-x 

# Hydrometallurgical Extraction of Zinc and Copper – A $^{57}$Fe-Mössbauer and XRD Approach

A. F. MULABA-BAFUBIANDI[1,*,★] and F. B. WAANDERS[2]

[1]*Extraction Metallurgy Department, Faculty of Engineering, Technikon Witwatersrand, P.O. Box 526, Wits 2050, South Africa; e-mail: antoinemulaba@hotmail.com*

[2]*School of Chemical and Minerals Engineering, North-West University (Potchefstroom campus), Potchefstroom, 2531, South Africa; e-mail: chifbw@puk.ac.za*

**Abstract.** The most commonly used route in the hydrometallurgical extraction of zinc and copper from a sulphide ore is the concentrate–roast–leach–electro winning process. In the present investigation a zinc–copper ore from the Maranda mine, located in the Murchison Greenstone Belt, South Africa, containing sphalerite (ZnS) and chalcopyrite ($CuFeS_2$), was studied. The $^{57}$Fe-Mössbauer spectrum of the concentrate yielded pyrite, chalcopyrite and clinochlore, consistent with XRD data. Optimal roasting conditions were found to be 900°C for 3 h and the calcine produced contained according to X-ray diffractometry equal amounts of franklinite ($ZnFe_2O_4$) and zinc oxide (ZnO) and half the amount of willemite ($Zn_2SiO_4$). The Mössbauer spectrum showed predominantly franklinite (59%), hematite (6%) and other Zn- or Cu-depleted ferrites (35%). The latter could not be detected by XRD analyses as peak overlapping with other species occurred. Leaching was done with HCl, $H_2SO_4$ and $HNO_3$, to determine which process would result in maximum recovery of Zn and Cu. More than 80% of both were recovered by using either one of the three techniques. From the residue of the leaching, the Fe-compounds were precipitated and <1% of the Zn and Cu was not recovered.

**Key Words:** $^{57}$Fe-Mössbauer spectroscopy, leaching, roasting, sulphide ore, XRD-analyses.

## 1. Introduction

Zinc in nature occurs mainly in sulphide ores as sphalerite (ZnS), but mostly other minerals such as pyrite ($FeS_2$), magnetite ($Fe_3O_4$), chalcopyrite ($CuFeS_2$), pyrrhotite ($Fe_{1-x}S$) and gangue materials accompany the ore [1]. The iron present in the ore makes recovery of the valuable zinc more difficult as zinc ferrites are being co-produced during high temperature roasting and their insolubility in relative dilute sulphuric acid results in a significant amount of zinc remaining in the residue and thus a source of metal loss. However the most commonly used

* Author for correspondence.

★ From January 2005 the Technikon Witwatersrand (TWR) will complete the merger with the Rand Afrikaanse Universiteit (RAU) to form the University of Johannesburg (UJ).

route in the hydrometallurgical extraction of zinc and copper from such sulphide ores is the concentrate–roast–leach–electro winning process. Firstly the ore is concentrated by means of flotation to eliminate all the gangue materials. The concentrate is then roasted to produce a calcine that will be leachable to extract the zinc and copper and finally electro winning is used to obtain the pure metal.

The first step in the process of beneficiation of the run-of-mine (ROM) ore is flotation, which is a technique that permits the recovery of the valuable ore minerals in the concentrate, from solution through the introduction of surface-active agents such as xanthate [2]. After flotation, roasting follows and various authors [3–7] have studied the roasting process of sulphide ores where the Fe is liberated and then precipitated to ensure maximum zinc and copper recovery. The calcine which is formed during roasting typically contains zinc and copper in the form of oxides (ZnO, CuO), sulphates ($ZnSO_4$, $CuSO_4$) and ferrites (Zn, $Cu_{1-x}M_x$)$OFe_2O_3$ or (Zn, Cu)$Fe_2O_4$ and thus the roasting process ensures that compounds suitable for further leaching are produced. Leaching, the final step to produce a solution containing the valuable metals to be extracted by electro winning, is described as a heterogeneous (solid/liquid) reaction and is normally conducted by means of adding an acid to the roast product, which can be a complex process [8] involving various chemical reactions to occur in parallel or in series. In the present investigation a zinc–copper ore from the Maranda mine, located in the Murchison Greenstone Belt in South Africa was studied. The main ore minerals, as determined by X-ray diffractometry conducted at the mine, were sphalerite (ZnS, 40%), pyrite ($FeS_2$, 26%), magnetite ($Fe_3O_4$, 8%), chalcopyrite ($CuFeS_2$, 8%) and pyrrhotite ($Fe_{1-x}S$, 4%), with the gangue material (14%) being clinochlore $(Mg, Fe)_6(Si, Al)_4O_{10}(OH)_8$.

## 2. Experimental

The as-received ore from the mine was ground to a particle size of 80 μm. The composition of the ore was confirmed from both mineralogical and chemical analyses. For the flotation of the ground ore, xanthate SNPX was used as a collector and dowfroth 400 was used as a frother. After bulk flotation of the ore most of the gangue material was removed and the concentrate was subjected to X-ray diffractometry and $^{57}$Fe-Mössbauer spectroscopy. Consequently, roasting of the concentrate was done on 25-g batch samples, which were put in a ceramic crucible lined with silica and then put into a laboratory furnace. The roasting temperatures varied from 700 to 1000°C and the duration of the roasting varied between 2 and 4 h until optimum roasting conditions were found to produce a calcine suitable for leaching. After completion of each batch roasting process the samples were again subjected to XRD and Mössbauer spectroscopy.

Neutral agitation leaching of the calcine, mixed with HCl, $HNO_3$ or $H_2SO_4$ was conducted in a 1-L laboratory beaker to determine the optimum leaching reagent and leach temperature. The density of the slurry to be leached was kept at 20% solids and the temperature was altered to determine which process conditions would result in the maximum recovery of the valuable metals. The residue of the optimum neutral leach process was finally subjected to hot acid leaching to precipitate the Fe and recover the remaining Zn and Cu. XRD and AAS analyses were conducted to determine the species whilst Mössbauer spectroscopy was performed to determine the Fe-containing species in these samples. The residual sulphur was determined by means of gravimetric measurements, according to the procedures described by Lenahan and Murray-Smith [9].

The XRD-analyses were conducted with a Philips PW 1830 diffractometer with a Cu-anode operating at a generator voltage of 40 KV and a current of 20 mA with the goniometer $2\theta$ values varied from 10° to 70° at a scan rate of 1.0 s/step. All Mössbauer spectra were obtained with the aid of an Austin Associates K3 Mössbauer spectrometer, capable of operating in conventional constant acceleration mode using a backscatter-type gas-flow detector. A 10-mCi $^{57}Co$ source, plated into a Rh-foil, was used to produce the $\gamma$-rays. The samples were analysed at room temperature and data collected in a multi-channel analyser (MCA) to obtain a spectrum of count rate against source velocity. A least-squares fitting program was used and by superimposing Lorentzian line shapes, the isomer shifts, quadrupole splitting and/or hyperfine magnetic field of each constituent was determined with reference to the centroid of the spectrum of a standard $\alpha$-iron foil at room temperature with the aid of the fitting program NORMOS-90. For the determination of the hyperfine magnetic field of the calcine, a hyperfine field distribution was used, resulting in an average field strength for the mixed ferrites. The amount of each constituent present in a sample was determined from the areas under the relevant peaks.

## 3. Results and discussion

Optimum flotation conditions were achieved within 10 min and the Mössbauer spectrum for the concentrate yielded pyrite (60%), chalcopyrite (21%) and clinochlore (19%). The pyrite and chalcopyrite were also observed in the XRD spectrum but due to the fact that the clinochlore cannot be assigned unambiguously from XRD data, it is not reported in the XRD spectrum (see Figures 1 and 2 and Table I). From the data of the original ore the clinochlore presence had been mentioned and from Figure 1, some lines ($2\theta \approx$ 12.6°, 18.9° and 25.3°) have not been specifically identified, where it is not clear whether they belong to clinochlore, although in agreement with the data of the original ore and thus the $Fe^{2+}$ and $Fe^{3+}$ identified in the Mössbauer spectrum of the concentrate will be attributed to clinochlore.

The concentrate was subsequently roasted at temperatures of 700, 800 and 900°C, each for periods of 2, 3 and 4 h respectively. The initial amount of sulphur (17%) was reduced to 1% after 4 h (see also Figure 3). Optimizing the

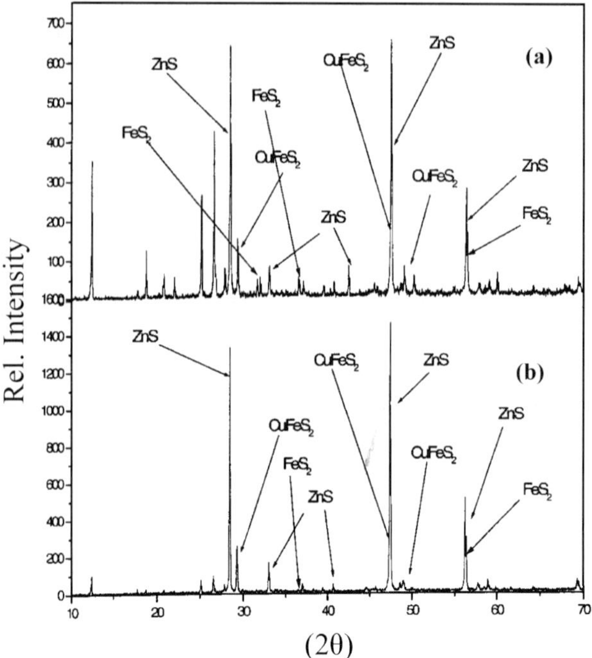


*Figure 1.* The XRD spectra of the (a) ROM ore and (b) the concentrate.

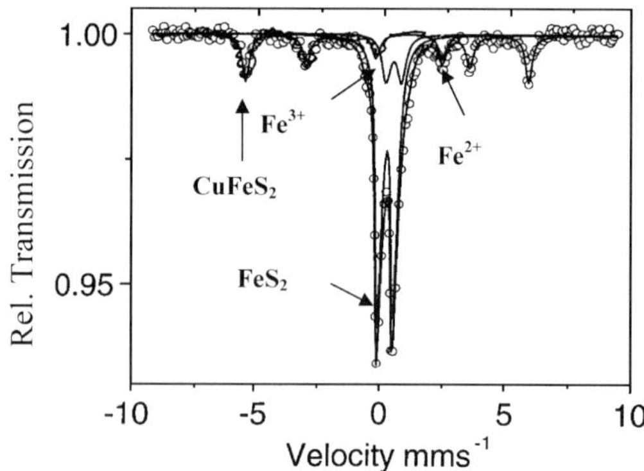


*Figure 2.* The Mössbauer spectrum of the concentrate with $Fe^{2,3+}$ attributed to clinochlore (gangue).

roasting conditions required that samples were subjected to different temperatures for different durations to see if time had any effect on the resulting calcine.

When one uses the combined requirement (low sulphur, high zinc oxide and low ferrites contents, and shorter roasting duration) as the criteria for the optimum roasting condition, it was found that 800°C, 3 h gave the best combination for low ferrites (36%) and low sulphur (2.92%), while 900°C 3 h could be chosen for its combination of low ferrites (34%) and high zinc oxide (66%) contents. As

*Table I.* Hyperfine interaction Mössbauer parameters of Fe-components found in this investigation

| Sample | Component | IS, $mms^{-1}$ (± 0.02) | QS, $mms^{-1}$ (± 0.02) | H Tesla (± 0.5) | Relative intensity (%) |
|---|---|---|---|---|---|
| Concentrate | Chalcopyrite | 0.25 | − 0.01 | 34.9 | 21 ± 2 |
| | Pyrite | 0.30 | 0.62 | | 60 ± 2 |
| | Clinochlore $Fe^{2+}$ | 0.99 | 2.64 | | 6 ± 1 |
| | $Fe^{3+}$ | 0.40 | 0.61 | | 13 ± 2 |
| Roast at 900°C for 3 h | Franklinite | 0.33 | 0.51 | | 59 ± 2 |
| | Mixed ferrites | 0.35 | − 0.02 | 50.0 | 35 ± 2 |
| | Hematite | 0.36 | − 0.17 | 51.0 | 6 ± 2 |
| Leach precipitate | Oxihydroxide | 0.37 | 0.56 | | 67 ± 2 |
| | Unknown specie | 0.37 | 0.96 | | 33 ± 2 |

Note: IS = Isomer shift, QS = Quadrupole splitting and H = Hyperfine magnetic field strength.

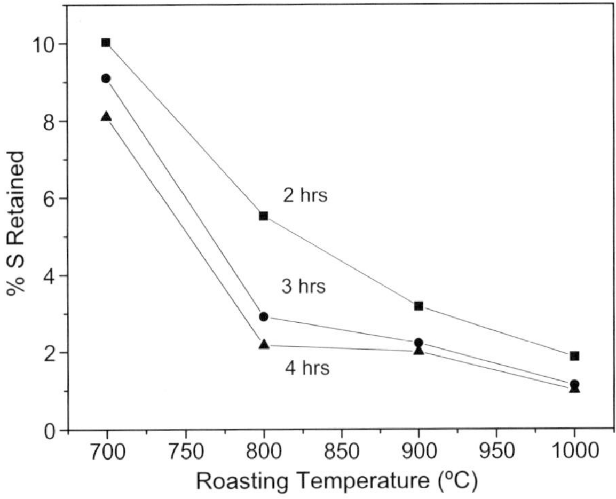


*Figure 3.* Graph showing %S retained at varying roasting conditions.

the roasting process is conducted for a subsequent leaching, one would want to dissolve at room temperature as much value as possible as a hot leaching would only escalate the operating costs.

The optimum roasting conditions were thus found to be 900°C for 3 h, which in the present investigation was higher than expected because the laboratory furnace used, had a poor supply of oxygen to blow out the sulphur to enhance the oxidation of the sulphide and thus the optimum temperature for sulphur removal could have been lower, as is the case in industry [8]. A relative good dead roasting was however obtained at these conditions with less than 3% sulphur remaining.

The XRD and Mössbauer spectra, obtained after roasting the concentrate at various temperatures, are shown in Figures 4 and 5. The calcine produced at the optimal conditions (i.e., minimum sulphur, maximum final metal recovery) contained equal amounts of franklinite ($ZnFe_2O_4$) and zinc oxide (ZnO) and half the

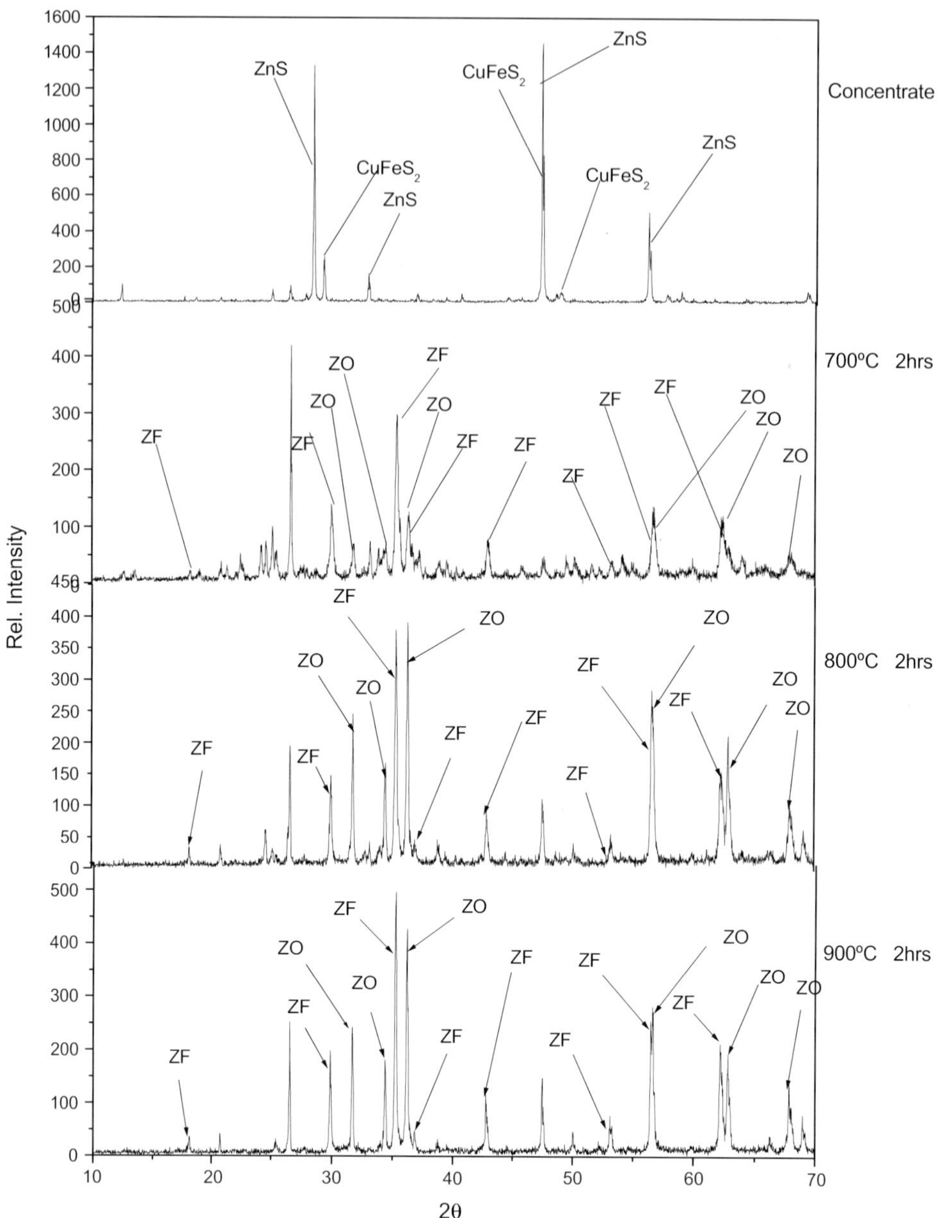


*Figure 4.* XRD-spectra showing the effect of roasting temperature for a specific time on the formation of zinc oxide and zinc ferrite. Note: *ZO* = Zinc Oxide, *ZF* = Zinc Ferrite.

amount of willemite ($Zn_2SiO_4$). The Mössbauer spectrum obtained for the samples roasted at the 900°C roasting temperature (see Figure 5) showed predominantly franklinite (59%), hematite (6%) and other Zn- or Cu-depleted ferrites (35%). The latter could not be detected with the aid of XRD analyses as the

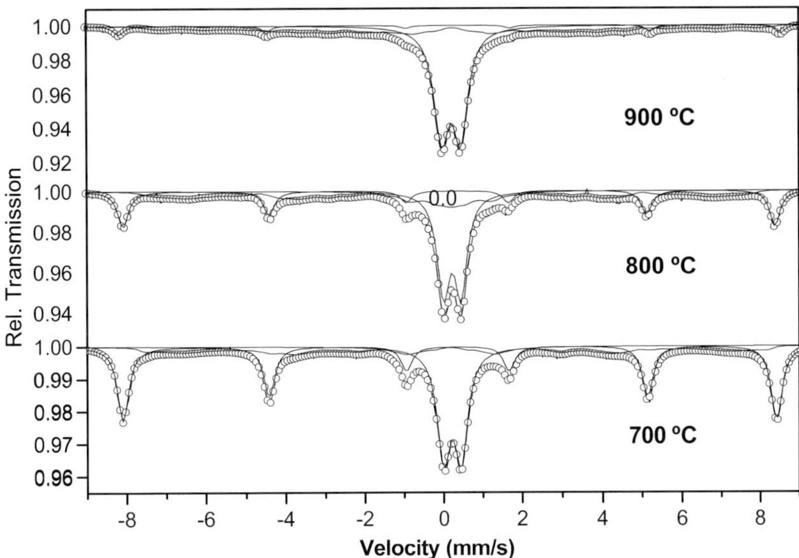


*Figure 5.* Mössbauer spectra of calcine roasted for 3 h. The overall fit to the spectrum is the *solid line* through the data points represented as *open circles*. The spectrum is fitted with a sextet representing a crystalline hematite phase, an intense central doublet representing franklinite and a distribution of sextets probably representative of other ferrite phases in the sample.

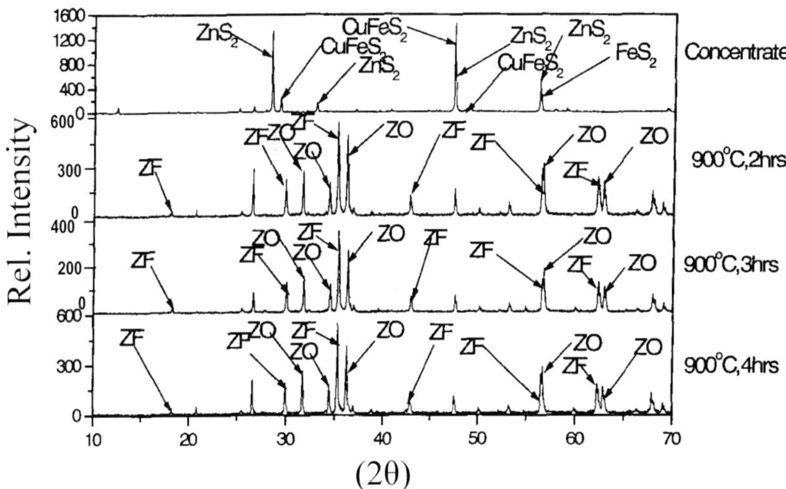


*Figure 6.* XRD-spectra showing the effect of roasting time for the 900°C temperature on the formation of zinc oxide and zinc ferrite. Optimum conditions were at 3 h roasting time. Note: *ZO* = Zinc Oxide, *ZF* = Zinc Ferrite.

peaks in the spectrum, Figure 6, overlap with those of the zinc ferrites and zinc oxides. The Mössbauer parameters are in accordance with those reported in literature [10] and are summarized in Table I together with the Mössbauer parameters found for the other components investigated in the present investigation.

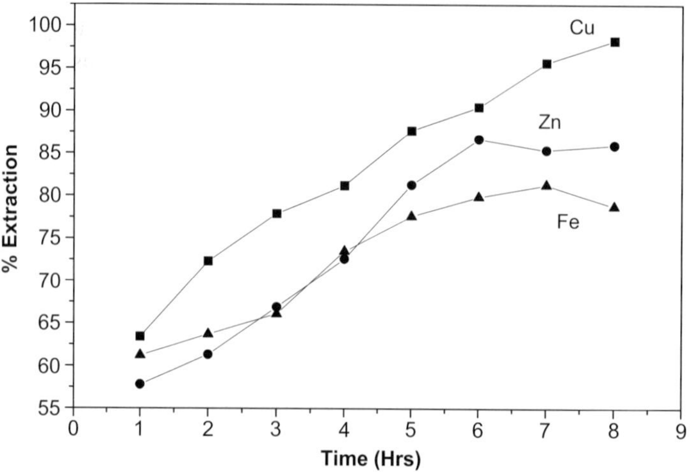


*Figure 7.* The amount of each metal extracted in $H_2SO_4$ at room temperature.

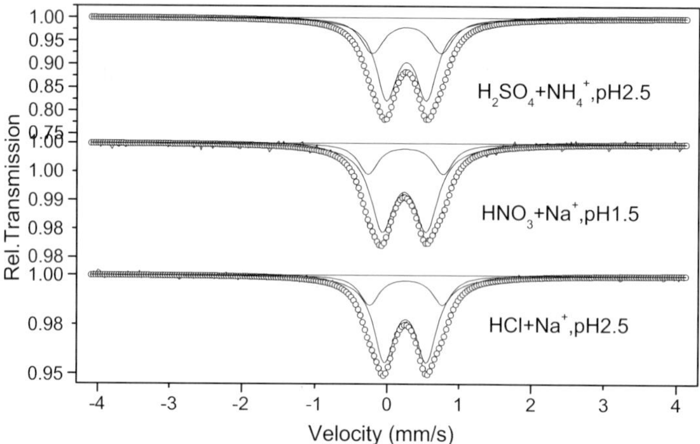


*Figure 8.* The Mössbauer spectra of the Fe-residue precipitated. The two *doublets* represent jarosite and an oxihidroxide respectively.

From the leaching experiments done with the three different acids to determine which process would result in maximum recovery of the valuable metals, it was found that almost all the Cu was leached within 8 h, whilst about 85% of the Zn was leached during the same period using either one of the three neutral acid leach techniques (see Figure 7 for the $H_2SO_4$ leaching results).

The residue from the neutral leach was submitted to a hot acid leaching to precipitate the Fe and recover the remaining Zn and Cu. Less than 1% of the metals were not leached in the final hot leach. Typical Mössbauer spectra for the Fe-products that were precipitated are given in Figure 8. From the XRD results the poorly crystalline or amorphous phases could not be identified and the Mössbauer parameters for the possible species precipitated, differed from jarosite, schwertmannite or ferrihydrite, typical minerals found to precipitate under plant conditions. Most probably some form or combinations of species of oxihydroxide, or even small super paramagnetic particles, were observed in the present investigation. To clarify the oxihydroxide species found, low temperature Mössbauer spectroscopy must be used, but unfortunately no such experimental setup was available during experimentation and further investigation has to be conducted.

## 4. Conclusion

From the sulphur tests the optimum roasting conditions were determined to be 900°C for 3 h and leaching in the three different acids yielded a recovery of more than 85% for the Zn with even higher recoveries for Cu. The exact phase composition of the precipitated Fe-containing residue could not be established and should be investigated further.

With a relative simple laboratory experimental set-up it is clear that Mössbauer spectroscopy, complemented with XRD-analyses can form the base for the identification of species present during the concentrate–roast–leach–electro winning process of a sulphide ore and further studies should be done to find alternative routes to obtain maximum recovery of valuable ores.

## Acknowledgements

One of the authors (A.F. M-B) is greatly indebted to the research committee of the Technikon Witwatersrand, Johannesburg, South Africa and to the National Research Foundation (NRF) for project funding. The experimental work was carried out by one of his students, F.N. Magagula and both authors want to thank her for the efforts she has put into the project.

## References

1. Hammerbeck E. C. I. and Mehliss A. T. M., In: Ore Deposits of the Republic of South Africa (eds.), *Department of Mining and the Geological Survey of South Africa, The Government Printer*, Pretoria, South Africa, 1976, pp. 461.
2. Habashi F., *Textbook of Hydrometallurgy*, Metallurgie Extractive Quebec, Quebec, Canada, 1999, p. 739.
3. Ozberk E. and Minto R., *Proceedings: Iron Control in Hydrometallurgy Symposium*, Toronto, Canada, 1986, pp. 12–54.
4. Stein J. Y. and Spink D. R., *Hydrometallurgy* **41** (1990), 156–161.
5. van Niekerk C. J. and Begley C. C., *J. S. Afr. Inst. Min. Metall.* **91** (1991), 233–248.
6. Laksham Y., Rathie V. I. N. and Monzyk B., *Iron Control and Disposal* (1992), 357–364.
7. Chen T. T. and Cabri L. J., *Can. Inst. Min. Metall.* (1993), 23.

8. Junea J. M., Singh S. and Bose D. K., *Hydrometallurgy* **41** (1996), 201–209.
9. Lenahan W. C. and de Murray-Smith R., *Assay and Analytical Practice in the South African Mining Industry*, 1986, 156–160.
10. Stevens J. G., Khasanov A. M., Miller J. W., Pollak H. and Li Z., In: Mössbauer Mineral Handbook, (eds.), *Mössbauer Effect Data Centre*, University of North Carolina, Asheville, USA, 1998, p. 527.

Hyperfine Interactions (2005) 161:43–53
DOI 10.1007/s10751-005-9190-4


# Magnetic and Mössbauer Studies of Quaternary Argentine Loessic Soils and Paleosols

R. C. MERCADER[1,*], F. R. SIVES[1], P. A. IMBELLONE[2] and R. E. VANDENBERGHE[3]

[1]*Departamento de Física, IFLP, Facultad de Ciencias Exactas, Universidad Nacional de La Plata, CC. 67, 1900, La Plata, Argentina; e-mail: mercader@fisica.unlp.edu.ar*

[2]*Instituto de Geomorfología y Suelos, Universidad Nacional de La Plata, 3N°584, 1900, La Plata, Argentina*

[3]*NUMAT, Department of Subatomic and Radiation Physics, Ghent University, Ghent, Belgium*

**Abstract.** This paper is a review of the current status about the remaining problems that are found in the investigation of the Quaternary Argentine soils and loessic sediments, and the way that Mössbauer studies can assist in solving them. There are two main types of investigations that make use of the magnetic response of the samples to correlate them with information gathered by other methods. On the one hand, there is the stratigraphic and chronological research, which is of importance from the geological and paleontological points of view. On the other hand, the paleoclimatic records, of significance toward a possible model of the past climate, are also studied because of their close relation to the sediments history. However, there is not yet a model that can tell the difference between the modifications due to the climatic conditions at the time when the soils were buried from processes that occurred after burial. Some examples are given that show that Mössbauer studies can be applied with a certain degree of success when cross-checked with magnetic measurements toward understanding the processes that occurred in alluvial B (paleosols) and C horizons (loess) from the eastern part of Buenos Aires Province. Although the application of Mössbauer studies to hydromorphic processes in soils is not straightforward, there are cases in which Mössbauer spectroscopy, if applied properly and correlated with other techniques, is able to characterize the type of iron oxides existing in the materials and thus assist theories about its origin and history.

**Key Words:** Argentina, hydromorphism, loess, magnetism, Mössbauer spectroscopy, paleosols, soils.

## 1. Introduction

Soils are natural deposits that are the result of the combined action of climate, living organisms, parental material, relief and time. The sediments of the quaternary plains have been formed by accretion of terrestrial silt carried and deposited by wind and water (loess) and can contain intercalated buried soils (paleosols). Paleosols are fossil records of past ecosystems considered as paleo-

* Author for correspondence.

climatic indicators, natural limits of stratigraphic sequences and carry support of ancient cultures.

The sedimentary magnetic properties that indicate the composition, concentration, and magnetic grain sizes of magnetic minerals, are parameters that are mainly used for two purposes; to give clues on paleoclimatic changes revealed by loess–paleosol sequences, and to correlate it with other mineralogical and grain size analyses parameters to establish more accurate stratigraphic and chronological records. The magnetic records worldwide are coincidental with respect to the existence of magnetic field polarity reversals from the past. These are taken as time marks for correlation with the fossil records found in the sediments and to establish patterns of paleoclimatic changes in different areas of the world.

Diverse magnetic responses of the loessic sediments have been reported in different continents. In particular, recent studies of the Argentine Pampas loess–paleosol sequences seem to confirm that they have a reverse trend to those which are well-established in China [1, 2]. The interest in these sequences is revealed by the intense research currently undertaken, which makes use of environmental magnetic approaches together with soil science techniques [3]. Since the magnetic response of iron oxides is forty times higher than all the other minerals together, many investigations use magnetic techniques and Mössbauer spectroscopy to gain information about the processes that the sediments have undergone over the ages.

Loess and reworked (loess-like) deposits in the northern hemisphere can be found in the mid-western plains of North America, in Alaska, in eastern parts of Europe, in the Russian–Siberian steppes, and in the Chinese plateaus [4]. These regions are concentrated along former periglacial zones.

The loess plateau of Argentina covers a large area in the central and northern part of the country and is the only loessic plain in the southern hemisphere. Loess–loessoid records of the last 5 to 3 Ma are recognized in southern Pampas and Undulating Pampas of the Province of Buenos Aires. Traditional ideas about the origin of the Pampean loess deposits show that the sources were the western mountain ranges of South America [5]. These first theories also accepted that loess has been transported by winds from distal areas mainly during cold and dry climatic periods. In the Pampean region, paleosols are very abundant and their stratigraphic records span the Pleistocene and Holocene ages. The loessic deposits exceed depths of 40 m in this area and are made up of superposed loess mantles of varying thickness, usually between 1 and 2 m.

In the literature, the hydromorphic changes produced in the paleosols are generally mentioned but not duly investigated. However, in the Pampean sediments, it has been found that hydromorphic processes were very important modifiers of the sediments. In particular, in the central part of the region, hydromorphic features have been found to increase with the age of the deposits [5].

In the broad range of Quaternary studies, one studied subject is the knowledge of the magnetic responses linked to the sedimentological and pedogenetic characteristics of materials both of present and of ancient systems.

In this work, we report preliminary studies that are currently conducted in different sections of the eastern Pampean region, taking into account not only the lithologic characteristics of the materials but also pedogenetic hydromorphic processes, which have been found to be responsible for the magnetic behavior of soils and sediments. We also discuss the likely information that Mössbauer spectroscopy can provide to better understand the paleoclimatic and stratigraphic records.

## 2. Magnetic studies of loessic soils and paleosols

The correlation of the magnetic records with the past climate has been found not to be universal [6]. Indeed, over the years the different investigations carried out across many regions of the world have shown that a correlation in one region is not universally applicable. Three main types of sequences have been found [7].

i) There is a typical sequence in which paleosols show a higher magnetic susceptibility than loesses. This is observed in the Russian plain, China and Central Asia. The increase in the magnetic response is mainly found in the clay fraction while the susceptibility of the coarse fraction is low and uniform. In addition, the features of the accompanying magnetite (which is unstable under chemical extraction treatments) suggest a pedogenic origin. In this type of sequence the magnetic records are generally correlated to paleoclimatic changes.

ii) There exists another type of sequence in which the correlation is reversed; loesses have a higher magnetic response, and the coarser loess fractions have even higher values. In this case the magnetic minerals (mainly magnetite) are stable when subjected to the same extraction treatments, suggesting the possibility that they are mainly made up of titanomagnetite. This type of sequence, found in Alaska and West Siberia, would not serve as a very good climatic proxy, instead, the magnetic record can yield information about the sedimentation processes.

iii) The third sequence found in Precaucassus loess exhibit a type with much more complex magnetic variations in which generally paleosols have a higher magnetic response than loesses, although both loess and paleosols have large variations in the susceptibility values. The current idea about this behavior associates the loess variation with sedimentary processes whereas those of the clay fraction of the paleosols are related to pedogenic processes. This type of sequence behaves like a mixture of the other two types and the very complex sequence is of limited value to understand the processes that have led to the current conditions.

The classic application of magnetic methods to stratigraphy has been recently extended to paleoclimatic, paleopedological and pedological applications. Very few magnetic studies in Argentina are found in the literature. Nabel *et al.* [8] have studied soil and underlying sedimentary sections located at different topographic positions in the Undulating Pampa. The magnetic susceptibility appears to be related to lithology, allowing the identification of some material and lithologic discontinuities. The maximum susceptibility values appear in both the B horizons of the present zonal soils as well as in the primary or reworked loess levels. In contrast, minimum values are associated with calcrete, paleosol, and hydromorphic horizons. The authors conclude that the superposition of effects on the magnetic signal precludes a direct interpretation of the magnetic results and that more complex studies and comprising more techniques would be required to arrive at more solid conclusions.

Contributions on magnetostatigraphy are more easily found and have been used as an age control of the sediments in which numerical values are absent [9]. The Bruhnes–Matuyama boundary is found in the eastern sections of the Province of Buenos Aires at depths between 7 and 10 m, and is related to paleopedological events. The paleoclimatic significance of paleosols and other indirect records have been used as clues for certain areas of the Pampean region, but more studies are necessary to establish a general paleoclimatic pattern for the entire Pampean plain (see review by Muhs and Zarate [10]).

## 3. Loess and its sedimentary characteristics

The Pampean region is a large sedimentary basin that has received sediments over the later part of the Cenozoic, with sediment thicknesses up to 120 m [11]. It is one of the more extended loessic plains in the world with a total area of 500,000 km$^2$ and the only one in the southern hemisphere. The pyroclastic volcanic nature of the Pampean loess makes it different in composition from the other three loessic plains since these are made up of granitic detritus from the direct action of the glaciers which was carried away by the wind to the deposit areas.

In 1957, Teruggi [12] already determined that the Pampean loess consists of sandy silt with abundant volcanic glass, and the initial ideas about its origin, sources of sediments and genetic processes was established. The traditional idea was that the Andean volcanism has controlled the geological and sedimentary evolution of the greater part of the Argentine territory. Quite a few Chilean volcanoes have spread volcanic materials to distances far away from the eruption centers, with well-recorded ashes that have been found as far as in Rio de Janeiro [13, 14].

Currently, it is considered that there are several sources of sediments and that the loess deposition is due to aeolian–fluvial processes that allow differentiating diverse Pampean areas with mineralogical and textural variations [15].

The mineralogical and volcanoclastic nature of the sediments manifests itself in the mineralogy of the sand fraction. Within the basic composition, both in the fine and coarse fractions, there are vertical discontinuities detectable due to the variation of the percentage of the component that allow separating depositional loess mantles from Pliocene to Holocene sediments [16]. There are also regional compositional variations linked to the mixture of sediments coming from the Brazilian shield in the northeastern area and by the Pampean ridges in the northwestern region [16].

## 4. Soils and paleosols

The predominant present soils of the Pampean plain belong to Mollisols, Alfisols, Vertisols and Entisols. The main pedological processes in paleosols are clay illuviation, vertilization, calcification and hydromorphisms. The loessic sequences exhibit several intercalated paleosols which are evidence that the loessic sedimentation has not been constant over the ages but has been formed by pulses of different intensity [17].

Across the eastern part of the Pampean region, there exists a great complexity in pedosedimentary processes because of the lithological variation, the small thickness of the loess mantles, and the superposition of pedological processes. In that sense, there is general agreement that many of the buried soils are welded soils without composite paleosols [5]. In these systems, it is very difficult to identify the pedological cycles as discrete units.

## 5. Hydromorphic processes in soils

The basic and applied knowledge about hydromorphic processes in Argentina is scarce. The majority of the studies refer to the interpretation about the genesis of the soils. Notwithstanding that the features were brought about by hydromorphic processes and are referred to both soils and paleosols, very little is known on the specific relationships between the present an past ecosystems. In the present soils the relationship with the climate and the relief is direct, but in paleosols the postdepositional processes can obscure this relationship even more so, if the intrazonal character of the aquic moisture regime is taken into account.

Hydromorphism in soils is developed by the influence of the climate, the relief and the anthropic activity. It is a bio-geochemical process that is produced in soils and sediments which are waterlogged over some time. During the waterlogging period, oxygen can be rapidly depleted when the micro-organism activity is high. After the $O_2$ has been depleted, the percentage of facultative and strictly anaerobic organisms increases, the $NO^{3-}$ ions disappear while the $Mn^{3+}$, $Fe^{3+}$, and $SO^{4-}$ ions are reduced sequentially. In this way, the reduced soils undergo denitrification, iron and manganese reduction, and methane production. These processes have been defined by the Soil Taxonomy key [18] as aquic conditions, which are defined by three parameters; redoximorphic features, saturation, and degree of reduction in a soil.

## 6. Correlations with the existing ratios of hematite and magnetite

According to several authors, e.g., [19], the history of the successions is closely related to the Fe-oxides and hydroxides. Magnetite and hematite are the main phases that can be used to study, in detail, the present and past processes the materials have undergone. The ratio of hydroxides to hematite and magnetite is a possible reliable parameter to be used as a paleoclimatic proxy. In this type of characterization, the traditional techniques used in this area demand not only physical but also chemical extractions, which yield results with relatively large error bars. However, magnetic susceptibility studies of total and grain size and magnetically separated fractions can provide important clues, which when crossed checked with Mössbauer spectroscopy, can lead to a better accuracy in a non-destructive way.

In particular, the type of magnetite present in soils, paleosols and loess has been used to better establish the type of processes along sequences of sediments from Cameroun and Brazil [19] and China [20].

## 7. Some examples

### 7.1. LOESSIC SOILS AND PALEOSOLS OF THE EASTERN BUENOS AIRES PROVINCE

Several investigations about the magnetic response of sediments of the Province of Buenos Aires can be found in the literature (see, e.g., [1] and references therein), which seem to be indicative of a contrary trend to that found for the Chinese plateau. That is, in the Chinese sediments, the loess susceptibility is lower than that of the soils or paleosols [20].

In spite that several authors have used Mössbauer spectroscopy to complement magnetic studies of soils and paleosediments [19–22], to our knowledge, Mössbauer spectroscopy had not been applied to Argentine loess–paleosol sequences earlier than Ref. [2]. Currently studies of several other sections in the region are undertaken.

Taking samples that have already been characterized by their mineralogical, stratigraphic, and field-susceptibility features, AC susceptibility and Mössbauer spectroscopy studies of sediments (bulk samples) and magnetically separated fractions were performed. Figure 1 shows results of loess and paleosol samples from one section of a quarry with coordinates: 34° 57′ S, 57° 52′ W. The presence of magnetite in the loess is clearly revealed by the Verwey

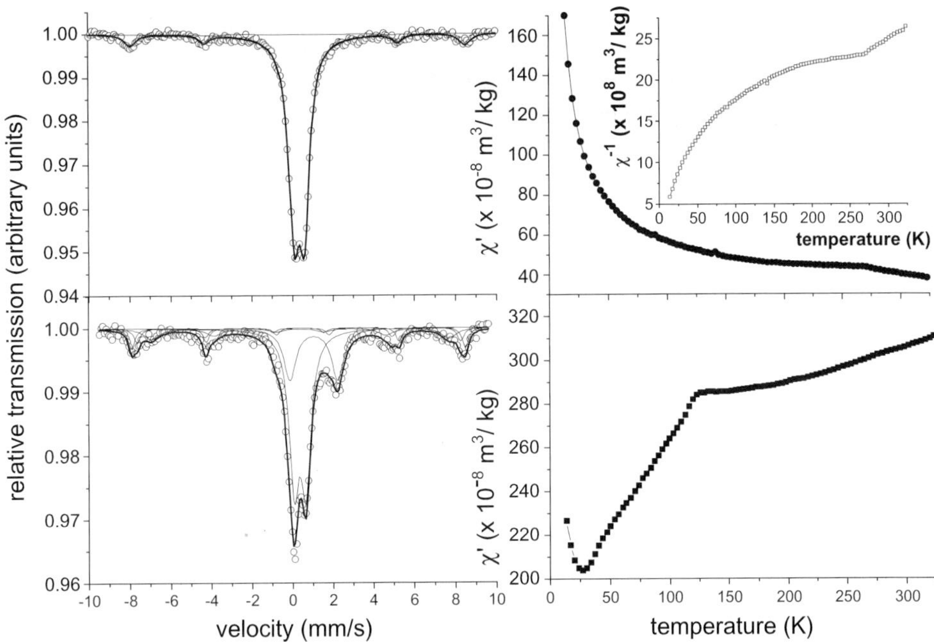


*Figure 1.* Mössbauer spectra (MS) at room temperature and AC susceptibility of two representative samples of Argentine Paleosols (*top figures*) and Loess (*bottom figures*). The inset in the top right figure is the thermal dependence of the inverse susceptibility.

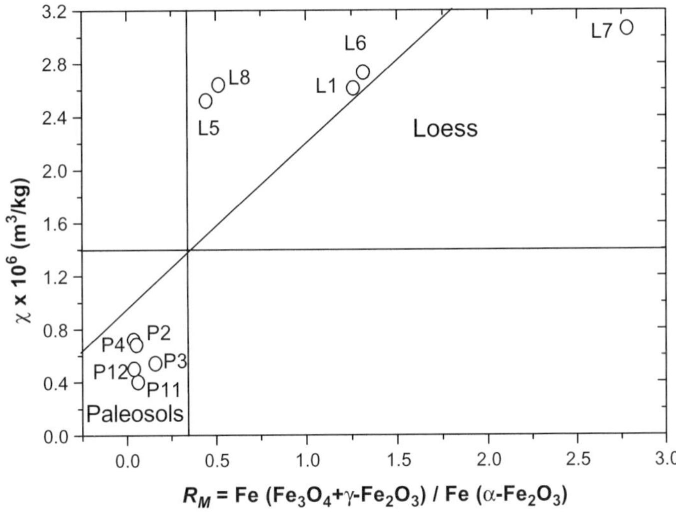


*Figure 2.* Magnetic susceptibility vs. Ratio of iron in magnetite (+maghemite) to iron in the hematite for different loess (*L*) and paleosol (*P*) bulk samples from a section in the north eastern area of Buenos Aires Province.

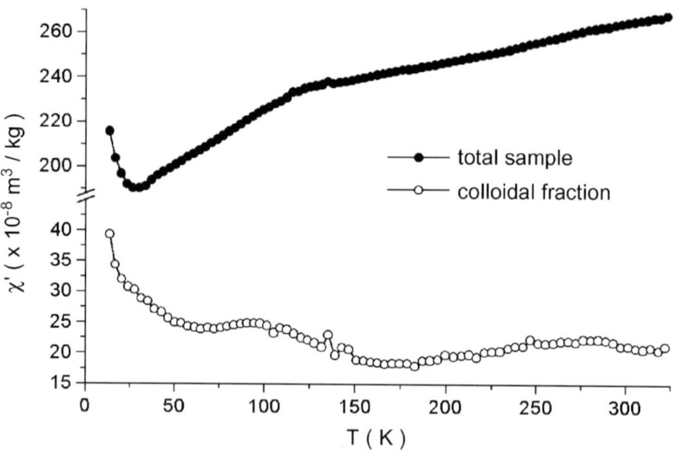


*Figure 3.* Thermal dependence of the magnetic susceptibility for the bulk (*solid points*) and the colloidal fraction (*open circles*) of one representative loess sample.

transition in the thermal dependence of the susceptibility, and although only room temperature Mössbauer spectra are shown, the difference between a paleosol sample (Figure 1, top) and a loess one (Figure 1, bottom) is already evident. The existence of magnetite is apparent in both the AC susceptibility and Mössbauer spectra, but hematite contributes $\approx 17\%$ and $\approx 12\%$ to the total area of the Mössbauer spectra of the paleosol and loess samples, respectively. In addition, the AC susceptibility results show a non-paramagnetic behavior, as can be seen in the inset at the top right of Figure 1. It displays the inverse susceptibility, denoting that the paleosol sample does not display any simple Curie–Weiss law.

When magnetic separations are performed on the samples, and the same studies as those described in Ref. [20] are carried out, valuable information can be obtained by plotting the amount of Fe in the various components (derived from the Mössbauer spectral areas) versus the measured susceptibility. This was performed on a selected, representative set of samples from the eastern Buenos Aires Province and the results shown in Figure 2. Similar to the results reported in Ref. [20], the plot of the relationship shows loess and paleosols samples in opposite regions of the graph but with a reverse trend.

In Figure 3 the thermal dependence of the AC susceptibility of one representative loess sample is shown, which allows for deducing that the main part of the magnetite is sedimentary and not of pedogenetic origin. Currently, no Mössbauer spectrum is available, which will most probably reveal more easily the existence of superparamagnetic magnetite in the colloidal fraction. This is an example where Mössbauer studies can be applied and in which the cross checking with magnetic measurements helps in the understanding of the system.

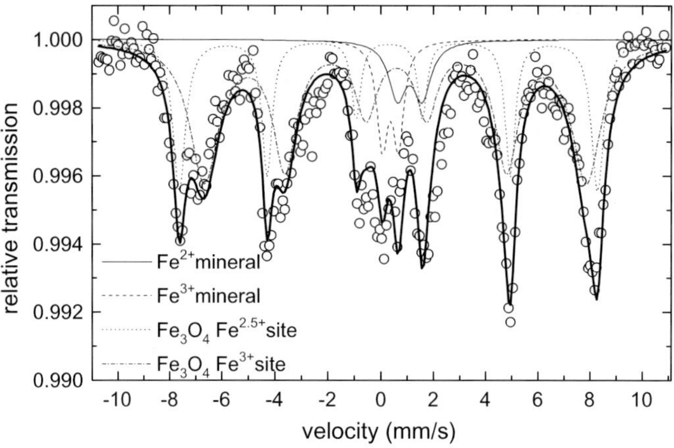


*Figure 4.* Room temperature Mössbauer spectrum of a magnetically enriched fraction of 88 to 125 μm size from a dune of the sandy Pampean area.

## 7.2. PRELIMINARY RESULTS OF MAGNETIC MINERALS FROM THE NORTHWESTERN BUENOS AIRES PROVINCE

In a project started recently, Mössbauer spectroscopy is used to characterize different magnetic minerals from sediments of the north western Buenos Aires Province. In Figure 4 the preliminary Mössbauer spectrum of a sample is shown, which has been sifted and magnetically enriched from a sample of sand dunes from the western part of Buenos Aires Province. The two doublets may belong to natural ilmenite, but with only this spectrum, it is not possible to discern whether the sextets correspond to titanomagnetites (Néel temperatures, $T_N$, of $\approx$300°C) or to a doped magnetite ($T_N \approx$ 500°C). Taking measurements at different temperatures and with better statistics, it may be possible to identify the mineral phases, which might yield information about the origin of the parental material.

## 7.3. HYDROMORPHIC SOIL STUDIES

Although in a very preliminary stage, Mössbauer studies have been applied to identify the type of iron oxides present in samples from soils belonging to the coastline of the La Plata River. This type of determination would be of great importance to understand the pedogenesis process. However, it should be noted that the extreme complexity of the system and severe limitations due to the overlapping signals in a Mössbauer spectrum arising from iron oxides and hydroxides with similar stoichiometric compositions, interpretations could be difficult, but these limitations, however, are not only restricted to Mössbauer spectroscopy, but also to several other techniques of common use in soil studies.

## 8. Final remarks

Since the magnetic, lithologic and pedological features of soils and paleosols are indirect records of climatic changes, they can be used as proxies for paleoclimatic studies. The lithological features are related to the quantity and size of magnetic minerals and the alteration products. In the second case they are also related to hydromorphic, melanization and calcification processes. In this extremely complex range of processes that are at play in the natural history of a sample, the general approach is to begin with a thorough analysis of the features of the Pampean sediments, which are better known and of the hydromorphic process, which is practically unknown. The study of the pedogenetic magnetite is thus a challenge.

Loessic sediments with a volcanoclastic composition are the parent material of the great majority of the soils and paleosols of the Pampean region. The presence of granitic minerals and metamorphic ones, originate in sedimentary contributions from different areas and carry scarce magnetic significant minerals.

In hydromorphic soils the temporal sequential analysis of the oxidation–reduction processes allow for the establishing of the evolution of those properties in the natural ecosystem, in addition to establishing relationships between the morphologic features of the soils, the chemical state and the behavior and interaction with the anthropic environment. In ancient soils it is only possible to speculate about the paleoclimatic conditions through the observation, identification and analysis of the features recorded in soils and sediments, supporting the findings with current conditions and research results. This is due to the fact that hydromorphism is a dynamic process of fast evolution, reversible and measurable in relation with respect to external present factors and sometimes modified by post depositional conditions.

The contribution of Mössbauer studies to the characterization of loessic sediments has already been demonstrated. The current investigations on the loess and loess-like Argentine sediments will provide clues to the paleoclimatic history if a strategy is devised capable of sorting out the local hydromorphic processes from those of the regional paleoclimatic features.

## Acknowledgements

This work has been partially supported by FWO-Flanders and SECTIP-Argentina under project FW/PA/02-EIII/003 and by PIP 2853, CONICET, Argentina. RCM is member of *Carrera del Investigador Científico* of CONICET. The authors thank Dr. J.C. Bidegain for providing some samples.

## References

1. Orgeira M. J., Walther A. M., Vázquez C. A., Di Tommaso I., Alonso S., Sherwood G., Yuguan H. and Vilas J. F. A., *J. S. Am. Earth. Sci.* **11** (1998), 561.
2. Terminiello L., Bidegain J. C., Rico Y. and Mercader R. C., *Hyperfine Interact.* **136** (2001), 97.
3. Deng C., Zhu R., Verosub K. L., Singer M. J. and Vidic N. J., *J. Geophys. Res.* **109** (2004), 1103.
4. Heller F. and Evans M. E., *Rev. Geophys.* **33** (1995), 211.
5. Imbellone P. A. and Teruggi M. E., *Quatern. Int.* **17** (1993), 49.
6. Maher B. A., *Palaeogeogr. Palaeocl.* **137** (1998), 25.
7. Virina E. I., Faustov S. S. and Heller F., *Phys. Chem. Earth A* **25** (2000), 475.
8. Nabel P. E., Morrás H. J. M., Petersen N. and Zech W., *J. S. Am. Earth Sci.* **12** (1999), 311.
9. Ruocco M., *Palaeogeogr. Palaeocl.* **72** (1989), 105.
10. Muhs D. and Zárate M., Late Quaternary eolian records of the Americas and their paleoclimatic significance, In: V. Markgraf (ed.), *Interhemispheric Climate Linkages*, Academic, New York, 2000, p. 183.
11. Hernández M., Private communication (2001).
12. Teruggi M. E., *J. Sedimentary Petrology* **27** (1957), 322.
13. Larsson W., *B. Geol. Inst. Univ. Upsala* **26** (1937), 27.
14. Imbellone P. A. and Camilión M. C., *Pedologie* **38** (1988), 155.
15. Zárate M. A., *Quaternary Sci. Rev.* **22** (2003), 1987.
16. Morrás H. and Delaune M., *Ciencia del Suelo* **3** (1983) 140 and references therein.
17. Teruggi M. E. and Imbellone P. A., *Ciencia del Suelo* **5** (1987) 175 (in Spanish).
18. Soil Survey Staff, *Keys to Soil Taxonomy*, Soil Conservation Service, USDA, 1999.
19. Vandenberghe R. E., De Grave E., Landuydt C. and Bowen L. H., *Hyperfine Interact.* **53** (1990), 175.
20. Vandenberghe R. E., Hus J. J. and De Grave E., *Hyperfine Interact.* **117** (1998), 359.
21. Eyre J. K. and Dickson D. P. E., *J.Geophys. Res.* **100** (1995), 17995.
22. Vandenberghe R. E., De Grave E., Hus J. J. and Han J., *Hyperfine Interact.* **70** (1992), 977.

Hyperfine Interactions (2005) 161:55–60
DOI 10.1007/s10751-005-9191-3


# Recovery of Heavy Minerals by Means of Ferrosilicon Dense Medium Separation Material

F. B. WAANDERS* and J. P. RABATHO
*School of Chemical and Minerals Engineering, North-West University, Potchefstroom 2520, South Africa; e-mail: chifbw@puk.ac.za*

**Abstract.** The diamond-bearing gravels found along South Africa's West Coast are being beneficiated by means of dense medium separation (DMS) to reclaim the alluvial diamonds. Granular ferrosilicon (Fe–Si) is used as the DMS material and at the end of each operation the Fe–Si is reclaimed from the process stream using a magnetic separator and is then recycled but losses of Fe–Si due to attrition, adhesion to the separation products, density changes and changes to the magnetic properties can occur. The gravel obtained from the mining operation is washed and screened before heavy mineral separation. The concentrate, tailings and Fe–Si samples were investigated by means of SEM and Mössbauer spectroscopy to determine where changes to the Fe–Si, or contamination could occur. The composition of the Fe–Si was determined to be Fe (76.1 at%), Si (20.3 at%), Mn (1.5 at%), Al (1.5 at%) and Cr (0.6 at%) resulting in a more or less ordered $DO_3$ phase with a calculated composition of $Fe_3Si$ for this Fe–Si, consistent with the Mössbauer results where two sextets with hyperfine magnetic fields of 18.6 T and 28.4 T were observed. After DMS, magnetite and ilmenite, the minerals found in the gravel, were still present in the concentrate. In the tailings virtually no magnetite or ilmenite was found and only a doublet, identified as an oxihydroxide, due to the abrasion of the Fe–Si, was found. After magnetic separation, to wash and clean the Fe–Si for re-use, it was found that magnetite and ilmenite were still present in the Fe–Si, which results in a change in density of the Fe–Si, resulting in a higher density and loss of valuable diamonds.

**Key Words:** dense medium separation, ferrosilicon, Mössbauer spectroscopy, SEM.

## 1. Introduction

The major diamond mining operations along South Africa's West Coast centres on the theory of the fluvial transport in the geological past of diamondiferous gravels from the rich kimberlites of the interior to the west coast through large drainage systems. The diamond-bearing gravels are the remains of the eroded tops of kimberlite formed within the hinterland of South Africa soon after the break-up of Gondwanaland [1]. The De Beers' Namaqualand Mines are situated along a 250 km stretch of the Atlantic coast of South Africa, with at least four major mining areas. Around 200 million carats of diamonds have been produced

* Author for correspondence.

*Figure 1.* SEM micrograph of the Fe–Si investigated in the present experiment.

with the highest concentrations of the diamonds located at the base of the old gravel beaches and river channels at depths of up to 100 meters below surface, although mining to date has only ventured some 40 ms down [2].

In the beneficiation process to reclaim the alluvial diamonds, ferrosilicon (Fe–Si) is being used as the dense medium separation (DMS) material. Two types of Fe–Si can be utilised in the DMS process, one being atomised rounded particles and the others milled angular particles. From Figure 1 it is clear that the Fe–Si used at the mine is angular, which could result in higher losses to occur due to abrasion [3]. At the end of each operation, the ferrosilicon is reclaimed from the process stream using a magnetic separator and is then subsequently recycled. Unfortunately, amongst others, losses due to attrition, adhesion to the separation products, density changes and changes to the magnetic properties of the Fe–Si occur, which cause poor recoveries in the mineral beneficiation process. For the present investigation five samples from the whole separation stream were received to determine were possible losses or contamination of the Fe–Si could occur.

## 2. Experimental

Five samples from the DMS circuit stream (Figure 2) were received to determine where possible losses or contamination of the Fe–Si could occur.

For the EDS analyses the powdered samples were embedded in a resin and the surface polished to ensure that no topographical contrast could lead to distorted

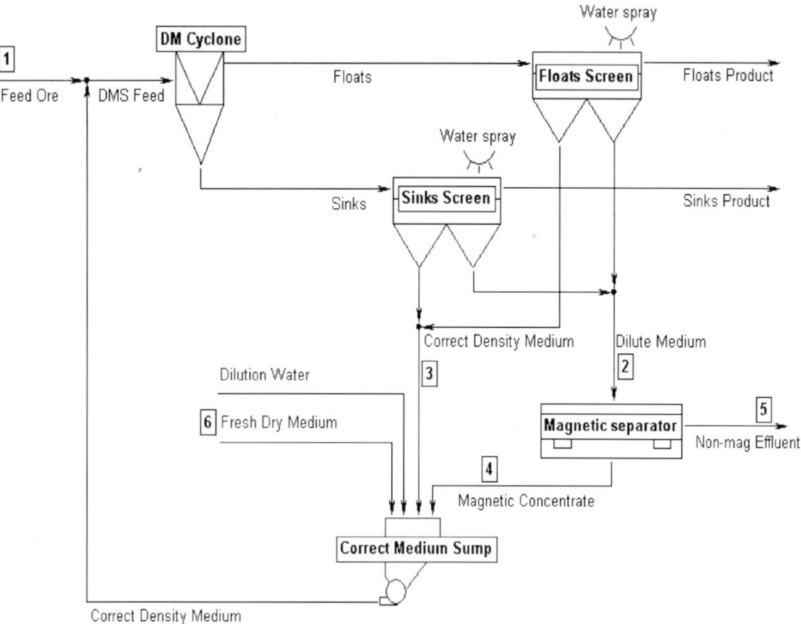


*Figure 2.* Typical DMS process used in the recovery of heavy minerals. The numbers indicate the positions where samples were taken for the present investigation.

values obtained. More than 100 analyses were done on each sample and averages were taken.

The average size distribution of the Fe–Si particles used, was measured with the aid of a Malvern 2000 Mastersizer and found to be $d_{0.5} = 45$ μm with the smallest particles having a size of 15 μm and the largest particles having a size of 115 μm. The density of the Fe–Si was determined to be 7.2 $g \cdot cm^{-3}$.

In the plant operations at the mine, the fresh Fe–Si is mixed with seawater to produce a slurry which has under pressure in the cyclone, a relative density of 3.2 $g \cdot cm^{-3}$. This material is then used as the DMS material, whereas the typical density of the diamonds to be recovered is 3.5 $g \cdot cm^{-3}$. The gravel, or feed ore, obtained from the mining operation is washed and screened before heavy mineral separation. In the magnetic separator, the magnetic materials (Fe–Si and other possible magnetic minerals present) are subsequently separated from non-magnetic minerals such as clay and quartz and the recovered Fe–Si recycled back into the system for further use.

All Mössbauer spectra were obtained with the aid of a Halder Mössbauer spectrometer, capable of operating in conventional constant acceleration mode using a proportional counter filled with Xe-gas to 2 atm. A 50 mCi $^{57}Co$ source, plated into a Rh-foil, was used to produce the $\gamma$-rays. The samples were analysed at room temperature and data collected in a multi-channel analyser (MCA) to obtain a spectrum of count rate against source velocity. A least-squares fitting

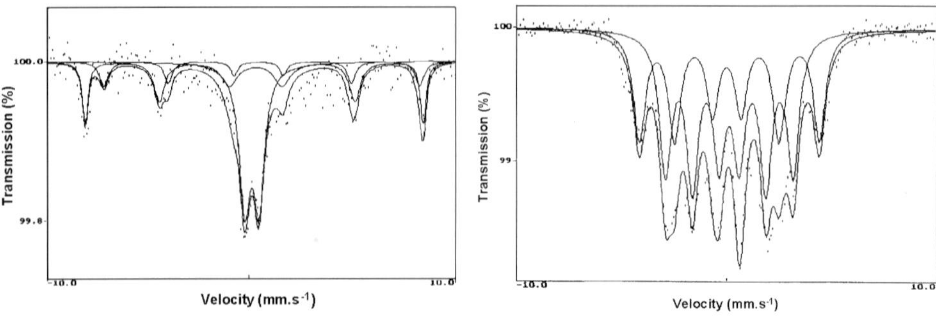


*Figure 3.* On the *left* the Mössbauer spectrum of the feed ore (containing presumably magnetite and ilmenite as Fe-bearing heavy minerals) and on the *right* the spectrum of the new, fresh Fe–Si with two sextets (magnetic hyperfine magnetic fields of 18.6 T and 26.4 T, respectively).

program was used and by superimposing Lorentzian line shapes, the isomer shifts, quadrupole splitting and/or hyperfine magnetic field of each constituent was determined with reference to the centroid of the spectrum of a standard α-iron foil at room temperature. The amount of each constituent present in a sample was determined from the areas under the relevant peaks.

A FEI Quanta 200 ESEM Scanning Electron Microscope fitted with an Oxford Inca 400 energy dispersive X-ray spectrometer (EDS) was used for the SEM analyses of the samples received.

## 3. Results and discussion

From the EDS analyses of the feed ore (sample position 1), the main non-ferrous minerals were found to be quartz and Ca-rich clays and the ferrous minerals were found to be magnetite ($Fe_3O_4$) and ilmenite ($FeTiO_2$), consistent with Mössbauer results (see Figure 3 left). The composition of the fresh Fe–Si (sample point 6) was determined from the EDS analyses to be Fe (76.1 at%), Si (20.3 at%), Mn (1.5 at%), Al (1.5 at%) and Cr (0.6 at%), resulting in a more or less ordered $DO_3$-phase with a calculated composition of $Fe_3Si$ for this Fe–Si, consistent with the results obtained for the Mössbauer spectrum where two sextets with hyperfine magnetic fields of 18.6 T and 26.4 T were observed (see Figure 3 right), smaller than the hyperfine magnetic fields of 20 T and 31 T of the ordered phase reported in [4]. A difficulty with the fitting of the sub-spectrum for the sextet with the hyperfine magnetic field of 26.4 T is the fact that it might be obscured by the presence of iron atoms in other environments when a possible eutectic forms.

After heavy medium separation, the magnetite and ilmenite originally present in the feed ore were still present in the concentrate as can be expected, due to their specific gravity of $>3.2\ g \cdot cm^{-3}$. In the tailings, or non-magnetic effluent (sample point 5), no magnetite or ilmenite was found and only a doublet, iden-

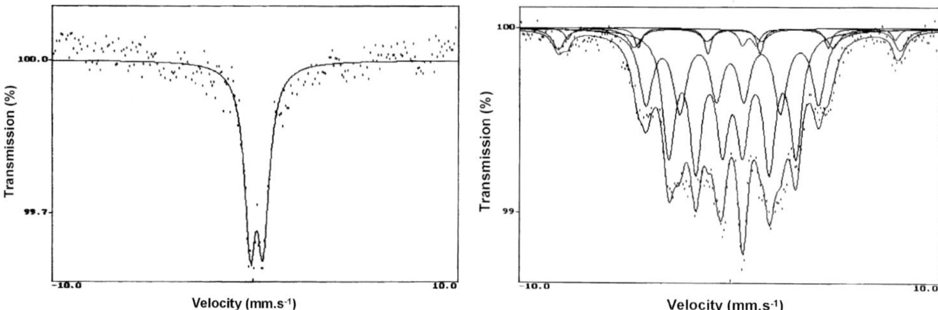


*Figure 4.* On the *left* the Mössbauer spectrum of the non-magnetic effluent indicating mainly the oxihydroxide due to abrasion of the Fe–Si and on the *right* the spectrum of the magnetic concentrate showing, in addition to the Fe–Si, the magnetite and ilmenite.

*Table I.* Mössbauer parameters of Fe-components found in this investigation

| Sample | Component | IS mms$^{-1}$ (±0.03) | QS mms$^{-1}$ (±0.03) | H Tesla (±0.3) | Relative intensity (%) |
|---|---|---|---|---|---|
| As-received Fe–Si | Sextet 1 | 0.16 | 0.12 | 26.4 | 43 |
| | Sextet 2 | 0.21 | −0.00 | 18.6 | 57 |
| Feed ore | Magnetite | | | | |
| | Sextet 1 | 0.38 | −0.17 | 49.9 | 37 |
| | Sextet 2 | 0.62 | 0.05 | 48.1 | 15 |
| | Ilmenite | 0.97 | 0.72 | – | 48 |
| Magnetic concentrate | Fe–Si | | | | |
| | Sextet 1 | 0.05 | 0.23 | 25.6 | 34 |
| | Sextet 2 | 0.23 | 0.02 | 18.6 | 52 |
| | Magnetite | | | | |
| | Sextet 1 | 0.20 | −0.19 | 50.0 | 8 |
| | Sextet 2 | 0.24 | −0.06 | 48.4 | 4 |
| | Ilmenite | 1.01 | 0.66 | – | 2 |
| Non-magnetic effluent | FeOOH | 0.28 | 0.58 | – | 100 |

Note: IS = Isomer shift relative to α-Fe, QS = Quadrupole splitting and H = Hyperfine magnetic field strength.

tified as an oxihydroxide due to the abrasion of the angular Fe–Si, was present (see Figure 4 left), which could lead in a loss of Fe–Si in the total circuit. The same mechanism was found to occur in an investigation of the use of Fe–Si in a hematite mine in South Africa [3]. After magnetic separation (sample point 4), when the Fe–Si has been washed and cleaned for re-use, it was found that magnetite and ilmenite were still present, although in lesser amounts (see Figure 4 right). This phenomena result in the DMS medium, Fe–Si, to change in density, resulting in a higher density of the separation material resulting in the loss of possible valuable minerals.

In Table I all the Mössbauer parameters, as determined in the present experiment are shown and the parameters as determined in the present investigation are in accordance to literature [6].

## 4. Conclusion

From the analyses of the samples it is clear that one mechanism of loss of dense medium material is due to abrasion. In a previous investigation [3] where the use of Fe–Si in a hematite mine was investigated it was found to be the biggest source of losses to occur. Furthermore valuable mineral losses could also occur at the mine due to the fact that the separation medium density changes as more and more of the heavy minerals such as magnetite and ilmenite stick to the Fe–Si that could not be separated from the Fe–Si during the magnetic separation part of the DMS circuit.

## Acknowledgements

The Kleinzee De Beers Diamond Mine is specially thanked for supplying the samples and Dr. L. R. Tiedt is thanked for his assistance with the SEM analyses.

## References

1. Hammerbeck E. C. I. and Mehliss A. T. M., In: Coetzee C. B. (ed.), *Ore Deposits of the Republic of South Africa*, Handbook 7, Department of Mining and the Geological Survey of South Africa, The Government printer, Pretoria, South Africa, 1976, pp. 461.
2. http://www.debeersgroup.com and personal communication.
3. Waanders F. B. and Mans A., *Hyperfine Interact.* **148–149** (2003), 325–329.
4. Stearns M. B., *Phys. Rev.* **129**(3) (1963), 1136–1144.
5. Nkosibomvu Z. L., Witcomb M. J., Cornish L. A. and Pollak H., *Hyperfine Interact.* **112**(1–4) (1998), 261–265.
6. Stevens J. G., Khasanov A. M., Miller J. W., Pollak H. and Li Z. (eds.), *Mössbauer Mineral Handbook*, Mössbauer Effect Data Centre, University of North Carolina, Asheville, USA, 1998, p. 527.

Hyperfine Interactions (2005) 161:61–68
DOI 10.1007/s10751-005-9192-2


# Structural and Electronic Properties Study of Colombian Aurifer Soils by Mössbauer Spectroscopy and X-ray Diffraction

H. BUSTOS RODRÍGUEZ[1,*], Y. ROJAS MARTÍNEZ[1], D. OYOLA LOZANO[1], G. A. PÉREZ ALCÁZAR[2], M. FAJARDO[2], J. MOJICA[3] and Y. J. C. MOLANO[4]
[1]*Departamento de Física, Universidad del Tolima, A. A. 546, Ibagué, Colombia; e-mail: hbustos@ut.edu.co*
[2]*Departamento de Física, Universidad del Valle, A. A. 25360, Cali, Colombia*
[3]*Departamento de Geología, Ingeominas Valle, Cali, Colombia*
[4]*Departamento de Geología, Universidad Nacional, Bogotá, Colombia*

**Abstract.** In this work a study on gold mineral samples is reported, using optical microscopy, X-ray diffraction (XRD) and Mössbauer spectroscopy (MS). The auriferous samples are from the El Diamante mine, located in Guachavez-Nariño (Colombia) and were prepared by means of polished thin sections. The petrography analysis registered the presence, in different percentages that depend on the sample, of pyrite, quartz, arsenopyirite, sphalerite, chalcopyrite and galena. The XRD analysis confirmed these findings through the calculated cell parameters. One typical Rietveld analysis showed the following weight percent of phases: 85.0% quartz, 14.5% pyrite and 0.5% sphalerite. In this sample, MS demonstrated the presence of two types of pyrite whose hyperfine parameters are $\delta_1 = 0.280 \pm 0.002$ mm/s and $\Delta_1 = 0.642 \pm 0.002$ mm/s, $\delta_2 = 0.379 \pm 0.002$ mm/s and $\Delta_2 = 0.613 \pm 0.002$ mm/s.

**Key Words:** arsenopyirite, chalcopyrite and galena, Mössbauer spectra, petrography analysis, pyrite, quartz, sphalerite, X-ray diffraction.

## 1. Introduction

The higher price of gold has contributed to a significant increase in the exploration and study of gold deposits. For most gold ores it is possible to recover up to 90% of gold using conventional techniques such as gravity, flotation and cyanidation. However, the use of these and other techniques depends on many factors, such as: the mineral phase to which the gold is associated (visible or invisible), the type of gold present, the grain size, etc. [1]. Then it is very important to obtain a good characterization of the mineral gold ore to determinate the recovering technique.

* Author for correspondence.

The intent of this paper is to complete the characterization of the Diamante gold mine, located in the municipality of Guachavez (Nariño)-Colombia. Previous studies showed that this is a precious metal deposit corresponding to a hydrothermal type composed principally by quartz and approximately 30% of sulfides, represented by pyrite, sphalerite and arsenopyirite [2]. The gold in particular is associated with pyrite and quartz and secondarily with arsenopyrite, sphalerite and galena. The composition of gold, established by EPMA, indicated 73% Au and 27% Ag (electrum type).

In the current work, the results of the Colombian auriferous soil study are discussed and complemented by using XRD and MS. Petrography analysis was made to each studied sample using optical microscopy.

## 2. Experimental

Thin polished sections of eight samples were prepared for the petrography analysis and were marked as samples number 805700, 805695, 805694, 805691, F2L1S1, PD14163, PD14163R and PD10114. The degree of reflectivity of each of the phases present in the sample was used as identification method. For the petrography analysis the OLYMPUS BH2 polarization microscopy was used with $\times 5$, $\times 10$, $\times 20$, $\times 50$, and $\times 100$ objectives. XRD pattern of each sample (polished thin sections) was taken using a RIGAKU Rint 2100 diffractometer, and the Material Analysis Using Diffraction (MAUD) program [3] was used for the Rietveld refinement. $^{57}$Fe Mössbauer spectroscopy analyses were used in the polished thin sections of the mineral samples. The obtained Mössbauer spectra were fitted using the MOSFIT program [4] and a $\alpha$-Fe foil as the calibration sample.

## 3. Results and discussion

### 3.1. PETROGRAPHY ANALYSIS RESULTS

In general, the petrography analysis of samples of the Diamante mine permits one to conclude that the principal phases are mainly the sulfides pyrite (cubic $FeS_2$), sphalerite (ZnS) and arsenopyrite (FeAsS). In the present investigation it was however found that pyrite is associated with arsenopyrite, and sometimes is associated with sphalerite, appearing as crystalline aggregates or isolated crystals. Arsenopyrite is commonly associated with the sphalerite and occasionally associated with pyrite. Sphalerite appears occasionally in isolated polycrystalline aggregates and normally associated with arsenopyrite and with pyrite. Chalcopyrite ($CuFeS_2$) is frequently associated with pyrite, arsenopyrite, sphalerite and freibergite [$(Ag, Cu, Fe)_{12}(Sb, As)_4S_{13}$]; sometimes found isolated in quartz. Galena (PbS) is associated with pyrite, arsenopyrite and sphalerite, and its occurrence is sporadic. Marcasite (orthorhombic $FeS_2$) is found generally

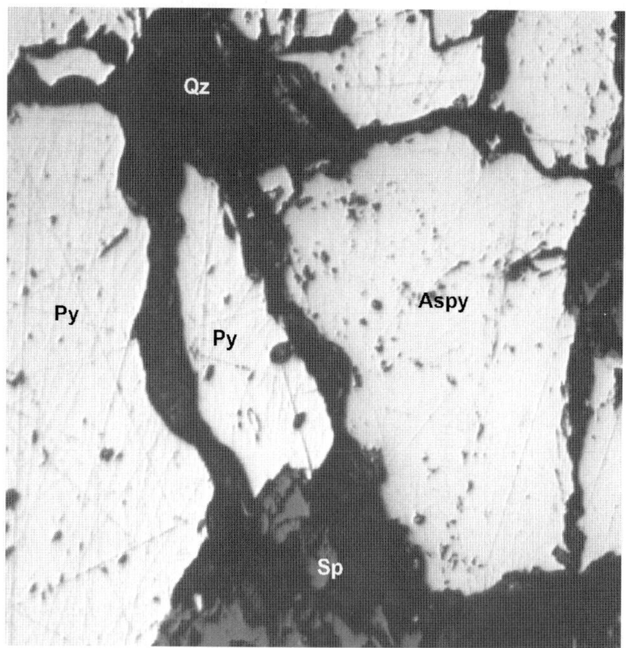


*Figure 1.* Photo micrograph of the 805691 mineral sample with optical camera, magnification ×20 (*Aspy*, arsenopyrite; *Py*, pyrite; *Qz*, quartz; *Sp*, sphalerite).

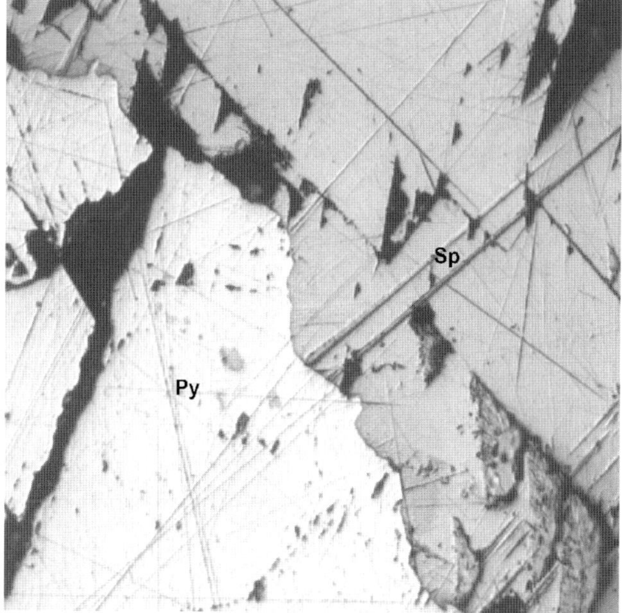


*Figure 2.* Photo micrograph of the F2L1S1 mineral sample with optical camera, magnification ×20 (*Sp*, sphalerite; *Py*, pyrite).

*Table I.* Results of the laboratory analysis for gold and constituent metals of the intact vein [2]

| Sample | Elements as determined by phantom atomic adsorption photometry | | | | | | | | | | | | | |
|---|---|---|---|---|---|---|---|---|---|---|---|---|---|---|
| | Au (ppm) | Ag (ppm) | Zn (%) | Fe (%) | Cu (%) | Pb (%) | As (%) | Mn (ppm) | Ni (ppm) | Mg (ppm) | Ca (ppm) | $SiO_2$ (%) | S (%) | Al (%) |
| 805691 | 11 | 80 | 4.70 | 27.1 | 0.21 | 0.40 | 6.70 | 435 | 55 | 190 | 30 | 29.8 | 28.2 | 0.38 |
| 805694 | 12 | 105 | 7.65 | 24.7 | 0.21 | 0.24 | 12.3 | 830 | 20 | 35 | 30 | 79.4 | 25.6 | 0.10 |
| 805695 | 5 | 500 | 5.81 | 8.92 | 1.92 | 0.42 | 1.47 | 500 | 20 | 40 | 20 | 74.7 | 10.2 | 0.05 |
| 805700 | 13 | 62 | 5.56 | 25.3 | 0.16 | 0.35 | 3.46 | 445 | 32 | 525 | 30 | 32.0 | 28.6 | 0.74 |

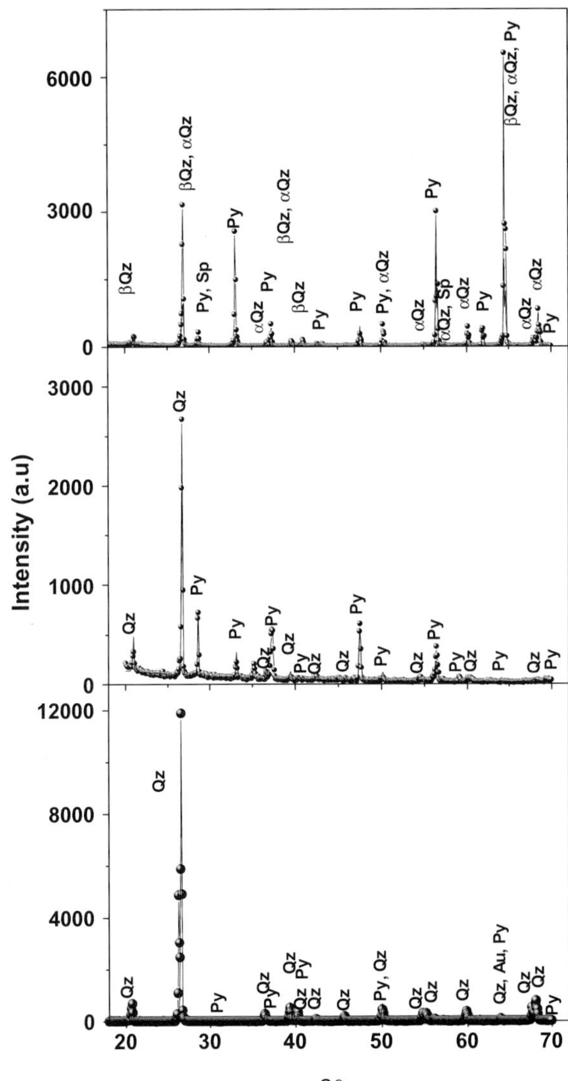


*Figure 3.* X-ray diffraction patterns of the 805694, 805691 and 805700 mineral samples.

included in pyrite, but its occurrence is very rare. It is generally associated with chalcopyrite, in accordance with the reported literature [5–7]. Typical microphotographs are shown in Figures 1 and 2, in which the samples 805691 and F2L1S1 are illustrated, respectively. It can be noted the association of pyrite with arsenopyrite, sphalerite and quartz in Figure 1 and pyrite with sphalerite in Figure 2.

Table I shows the chemical analysis conducted by atomic absorption of the samples 805691, 805694, 805695 and 805700. It can be noted that Au and Ag are obtained in some parts per million proving that the obtained gold in the electrum type.

*Table II.* Rietveld-refined lattice parameters, grain sizes, volume and weights of all the studied samples using the MAUD program

| Sample | Phases | Lattice parameters | | | Grain size (Å) | Fraction | |
|---|---|---|---|---|---|---|---|
| | | *a* (Å) | *b* (Å) | *c* (Å) | | Volume | Weight |
| 805694 | αQz | 4.910 | | 5.371 | 816.071 | 0.242 | 0.403 |
| | Py | 5.411 | | | 751.080 | 0.144 | 0.187 |
| | Sp | 5.423 | | | 499.619 | 0.004 | 0.004 |
| | βQz | 5.124 | | 5.096 | 497.351 | 0.611 | 0.407 |
| 805691 | Qz | 4.897 | | 5.411 | 501.318 | 0.806 | 0.686 |
| | Py | 5.400 | | | 509.294 | 0.194 | 0.314 |
| 805700 | Py | 5.429 | | | 3537.36 | 0.931 | 0.282 |
| | Qz | 4.919 | | 5.411 | 500.021 | 0.063 | 0.711 |
| | Au | 4.098 | | | 500.001 | 0.006 | 0.007 |

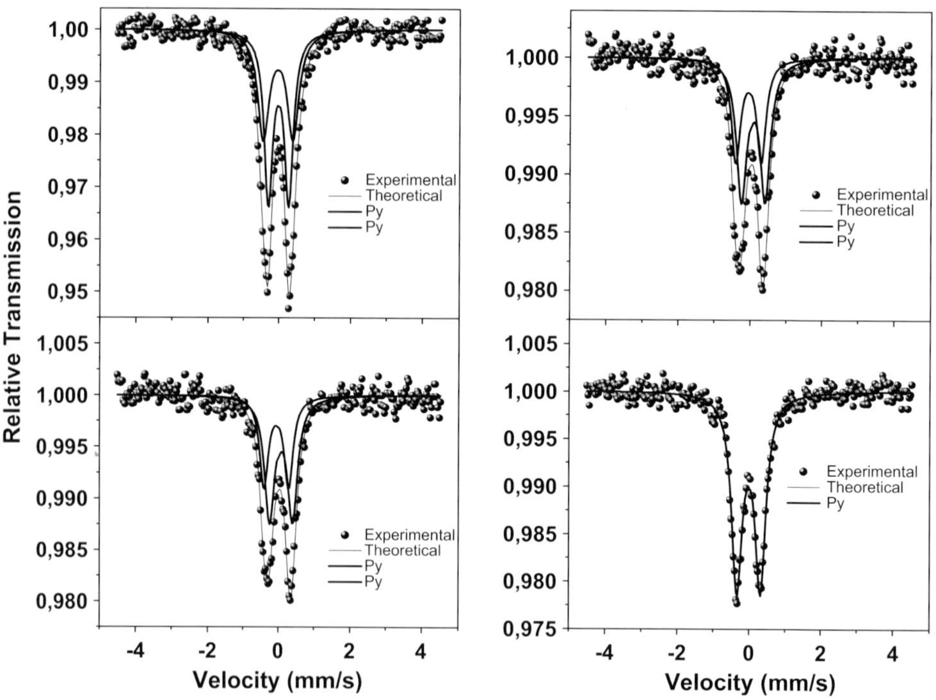


*Figure 4.* Mössbauer spectra of the 805694, 805691, 805700 and F2L1S1 mineral samples.

## 3.2. XDR RESULTS

Figure 3 shows the XRD patterns obtained for the 805694, 805691 and 805700 samples. From the Rietveld refinement conducted with the MAUD program α-quartz, β-quartz, pyrite and sphalerite were observed in the first sample, quartz and pyrite in the second and pyrite, quartz and gold in the third sample. A

*Table III.* Hyperfine parameters of the 805691, 805694, 805695, 805696 805700, PD1463 and F1L1S1 mineral samples obtained by using the MOSFIT program

| Sample | $\delta$ (mm s$^{-1}$) ± 0.002 | $\Gamma$ (mm s$^{-1}$) ± 0.002 | $\Delta$ (mm s$^{-1}$) ± 0.002 | $A$ (%) | Phase |
|---|---|---|---|---|---|
| 805688 | 0.144 | 0.181 | 0.635 | 53.94 | Py |
| | 0.563 | 0.164 | 0.649 | 40.06 | Py |
| 805691 | 0.307 | 0.154 | 0.560 | 57.11 | Py |
| | 0.315 | 0.214 | 0.866 | 42.89 | Py |
| 805694 | 0.280 | 0.170 | 0.642 | 59.23 | Py |
| | 0.379 | 0.177 | 0.613 | 40.77 | Py |
| 805695 | 0.317 | 0.147 | 0.540 | 54.18 | Py |
| | 0.299 | 0.195 | 0.799 | 45.82 | Py |
| 805696 | 0.144 | 0.177 | 0.644 | 50.58 | Py |
| | 0.552 | 0.173 | 0.635 | 49.42 | Py |
| 805700 | 0.368 | 0.171 | 0.617 | 60.34 | Py |
| | 0.234 | 0.151 | 0.665 | 39.66 | Py |
| PD1463 | 0.314 | 0.153 | 0.630 | 100.0 | Py |
| F2L1S1 | 0.32 | 0.180 | 0.613 | 100.0 | Py |

$\delta$ is the isomer shift relative to an iron foil at room temperature, $\Delta$ is the quadrupole splitting, $A$ is the relative spectral area, and $\Gamma$ is the line-width.

summary of the results obtained in these samples is presented in Table II. From this table it can be seen that the different postulated phases from the petrography analysis were also detected by XRD, except the arsenopyrite phase, showing that its presence is under the detection limit of the XRD technique. The obtained XRD parameters are in accordance with those reported in the literature [8].

### 3.3. MÖSSBAUER SPECTROSCOPY RESULTS

Figure 4 shows typical Mössbauer spectra for the 805694, 805691, 805700 and F2L1S1 mineral samples. The fits show that the only detected Fe phase is pyrite in all samples. It can thus be concluded that the other Fe phases detected by petrography and XRD are present but not in a large enough abundance to be detected by MS. The other phases do not content Fe. The obtained hyperfine parameters are shown in Table III. It can be noted that in some cases it was necessary to fit with two doublets, which correspond to two types of pyrite [9, 10].

## 4. Conclusion

From the present study on Colombian auriferous soils of the Diamante mine it can be concluded that the soils are rich in sulfides. Minerals such as pyrite, arsenopyrite, and sphalerite were the most frequently detected phases with pyrite commonly associated with the arsenopyrite and sphalerite.

## Acknowledgements

We are grateful to the Central Research Committee of the University of Tolima for financial support and to the Mössbauer Laboratory of the University of Valle where the experimental work was conducted.

## References

1. Harris D. C., *Miner. Depos.* **25** (1990), S3.
2. Molano J. C. and Mojica J. (eds.), Identificación de Sulfuros, sulfosales de plata y oro en la Mina el Diamante (Nariño)-Colombia, mediante análisis SEM-EDAX, Seminario Internacional, Ingeominas-Jica, Cali, 1998, p. 25.
3. Lutterotti L. and Scardi P. J., *Appl. Cristallogr.* **23** (1990), 246.
4. Varret F. and Teillet J., Unpublished MOSFIT program.
5. Friedl J., Wagner F. E., Sawicki J. A., Harris D. C., Mandarino, J. A. and Marion P., *Hyperfine Interact.* **70** (1992), 945.
6. Arehart G. B., Chryssulis S. L. and Kesler S. E., *Econ. Geol.* **88** (1993), 171.
7. Wu X., Delbove F. and Touray J. C., *Miner.Depos.* **25** (1990), S8–S12.
8. http://webmineral.com
9. McCammon C., *Mössbauer Spectroscopy of Minerals*, American Geophysical Union, 1995, p. 332.
10. Stevens J. G., *Mössbauer Mineral Handbook*, Asheville, North Carolina, 1998, p. 395.

Hyperfine Interactions (2005) 161:69–81
DOI 10.1007/s10751-005-9169-1 

# Vickers Microhardness and Hyperfine Magnetic Field Variations of Heat Treated Amorphous $Fe_{78}Si_9B_{13}$ Alloy Ribbons

A. CABRAL-PRIETO[1,*], F. GARCIA-SANTIBAÑEZ[2], A. LÓPEZ[2], R. LÓPEZ-CASTAÑARES[2] and O. OLEA CARDOSO[2]

[1]*Department of Chemistry, Instituto Nacional de Investigaciones Nucleares, Apartado Postal No. 18-1027, Col. Escandón, Del. Miguel Hidalgo, 11801, México, D.F., México; e-mail: acpr@nuclear.inin.mx*

[2]*Facultad de Ciencias, Universidad Autónoma del Estado de México, El Cerrillo Piedras Blancas, Toluca, Estado de México, México*

**Abstract.** Amorphous $Fe_{78}Si_9B_{13}$ alloy ribbons were heat treated between 296 and 763 K, using heating rates between 1 and 4.5 K/min. Whereas one ribbon partially crystallized at $T_x = 722$ K, the other one partially crystallized at $T_x = 763$ K. The partially crystallized ribbon at 722 K, heat treated using a triangular form for the heating and cooling rates, was substantially less fragile than the partially crystallized at 763 K where a tooth saw form for the heating and cooling rates was used. Vickers microhardness and hyperfine magnetic field values behaved almost concomitantly between 296 and 673 K. The Mössbauer spectral line widths of the heat-treated ribbons decreased continuously from 296 to 500 K, suggesting stress relief in this temperature range where the Vickers microhardness did not increase. At 523 K the line width decreased further but the microhardness increased substantially. After 523 K the line width behave in an oscillating form as well as the microhardness, indicating other structural changes in addition to the stress relief. Finally, positron lifetime data showed that both inner part and surface of $Fe_{78}Si_9B_{13}$ alloy ribbons were affected distinctly. Variations on the surface may be the cause of some of the high Vickers microhardness values measured in the amorphous state.

## 1. Introduction

The study of amorphous iron-based alloys has been extensive by now, and the amorphous to crystalline transition in bulk samples has been fairly well understood [1–5]. Mechanical and magnetic properties of these alloys change by thermal and other treatments on the amorphous or on the crystalline states, where the chemical composition, sample homogeneity and concentration of nucleation sites and defects determine the final properties of the alloy [6]. For instance, the Fe, Si and B contents in the $Fe_{78}Si_9B_{13}$ alloy is said to be within the

* Author for correspondence.

desired specifications as far as thermal stability is concerned. This alloy, however, may become completely brittle when crystallizes, limiting its applications in certain magnetic devices [7]. It might be possible that nanocrystalline $Fe_{78}Si_9B_{13}$ ribbons could preserve its original mechanical ductility, as in the amorphous state, due to the inverse Hall–Petch relation as previously shown by Jian-Min Li *et al.* [8] in $Fe_{78}Si_9B_{13}$ ribbons. In the nanocrystalline state of pure metals like Pd and Cu, this inverse Hall–Petch relation occurs within the crystal size range of 5 and 25 nm [9]. A mechanical softening that has not been elucidated yet [10, 11]. Jian-Min Li *et al.* [9] observed such a possible softening in $Fe_{78}Si_9B_{13}$ ribbons within the crystal size range of 25 and 140 nm, particularly between 70 and 65 nm. A study is required between the crystal size range of 5 and 25 nm, a range in which the pure metals soften systematically. It remains unclear if the inverse Hall–Petch relation indeed applies to metallic alloys. Particularly, the brittleness of $Fe_{78}Si_9B_{13}$ ribbons is largely due to the presence of crystalline phases like $\alpha$-(Fe–Si), $Fe_3B$ and $Fe_2B$ [1, 7], and other factors [8]. In the amorphous state, stressed, in-homogeneities in the sample, the concentration of nucleation sites and structural defects may be those factors affecting the physical properties of the amorphous $Fe_{78}Si_9B_{13}$ alloy. The interest here is to analyze the thermal induced mechanical effects, the hyperfine field variations, and the annihilation process of positrons in the amorphous state of $Fe_{78}Si_9B_{13}$ alloy ribbons, where atomic configuration variability, relaxation effects, creation and annihilation of defects are involved. In this sense, Vickers microhardness (VM), Mössbauer spectroscopy (MS) and Positron annihilation lifetime spectroscopy (PALS) data are presented.

## 2. Experimental section

Amorphous $Fe_{78}Si_9B_{13}$ alloy ribbons of 30 mm width and 25 μm thickness were used for this study [12], and several ribbons were thermally treated between 296 and 763 K. After annealing for 20 min at the selected temperature, five indentations were done on each side of the ribbons. The reported microhardness represents the average of ten measurements. The thermal treatments were carried out on a small furnace under vacuum ($4 \times 10^{-6}$ Torr). A sample, S1, 5 mm width and 10 mm large, was utilized for a first series of VM and MS measurements. Another sample, the S2 sample was used for a second series of VM measurements. This sample S2 was thermally-treated at the same time with a third sample, the S3, which was prepared as follows: A microdrop of an aqueous $^{22}$NaCl solution source, carrier-free, was put on the surface of a $Fe_{78}Si_9B_{13}$ alloy ribbon 5 mm width and 10 mm large. Water was evaporated using infrared light. The dried spot on the ribbon had an activity of 15 μCi. Two activity-free ribbons of the same aforementioned dimensions covered the ribbon with the spot and other ribbon was put on the other side of the activated ribbon to form a pile of four ribbons. This pile, covered with an Aluminum foil, was put into a Pyrex glass tube 40 mm large and 11 mm of internal diameter. The sample was evacuated once for 90 min using a vacuum system of $10^{-4}$ Torr. The evacuated tube with the sample in, was then sealed off. All thermal treatments were carried out on this evacuated tube-sample arrangement, sample S3. This sample S3 was used for PALS measurements. The heating rates used for sample S1 ranged from 1 to 3.7 K/min, and for samples S2 and S3 ranged from 1.97 to 4.47 K/min. Additionally, whereas for sample S1, a tooth saw shape was used for the heating and cooling rates, for samples S2 and S3 a triangular shape for the heating and cooling rates was used.

Microhardness measurements were performed on a microhardness Shimadzu tester using a Vickers diamond pyramid indentor. A 0.0025 N load was applied on the thermally treated samples for 10 s. The Mössbauer spectra were recorded using a conventional spectrometer operating in the constant accelerating mode with a $^{57}Co/Rh$ source. The reported isomer shifts are referred to metallic $\alpha$-Fe. A conventional fast-fast coincidence system with an instrumental resolution function of 0.36 ns and a short time window setting of 50 ns was used. The lifetime data of each spectrum contained more than $10^6$ counts.

## 3. Results and discussion

### 3.1. HEATING OF SAMPLES

Figure 1 shows the heating rates applied to samples S1, S2 and S3. These heating rates are low compared with those reported in the literature. For instance, J. Devuad-Rzepski *et al.* [1] utilized a heating rate of 80 K/min and a $T_x = 845$ K is reported for $Fe_{78}Si_9B_9$ ribbons. Xiangchen *et al.* [12] reports heating rates of 5, 10, 15, and 20 K/min and their corresponding $T_x$ values are 805, 817, 823 and 823 K, respectively. In the present work, the samples S1 and S2 crystallized at 763 and 722 K, respectively. The heating rates used for this purpose were 2.7 and 4.47 K/min, respectively. These temperatures values suggest that both samples were not completely crystallized since these temperatures are well below the 805 K that is obtained with a heating rate of 5 K/min. Figure 1 shows the thermal history, previously to the partial crystallization of samples S1, S2 and S3. Although the heating rates for sample S1 were slightly lower than for samples S2 and S3, a significant difference between their heating and cooling rates was involved between these samples, i.e., whereas a tooth saw shape was used in sample S1, a triangular form was used in samples S2 and S3. Selected X-ray diffraction (XRD) patterns, Figure 2, show that the heat-treated $Fe_{78}Si_9B_{13}$ alloy ribbons, between 293 and 673 K, remained amorphous. The recorded Mössbauer spectra of sample S1, after all these heat treatments, were also characteristic of the amorphous state, see later. These results agree with those reported by J. Y. Bang and R. Y. Lee [7], who worked at 673 K in $Fe_{78}Si_9B_{13}$. Additionally, S. Linderoth *et al.* [6] reports magnetic data on the as-quenched and heat treated

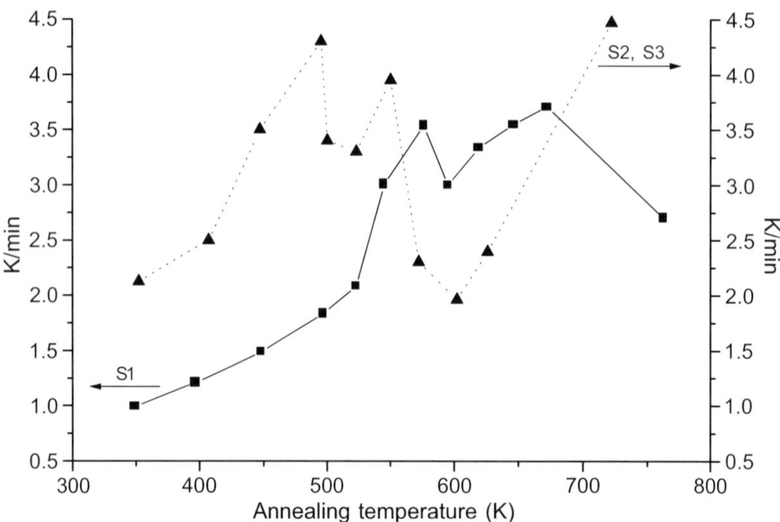


*Figure 1.* Heating rates applied to samples S1, S2 and S3. Whereas a *tooth saw form* was used for the heating and cooling rates of sample S1, a *triangular form* was used for the heating and cooling rates of samples S2 and S3.

amorphous $Fe_{81.5}Si_4B_{14.5}$ alloy at 673 K for 30 min and shows that their hysteresis loops remains the same with a coercive force of $H_c = 8 \times 10^{-7}$ T, suggesting that the amorphous state remains unaltered. Finally, after annealing samples S1, S2 and S3 at 763 and 722 K for 20 min at the heating rates indicated in Figure 1, respectively, the alloys became partially crystalline. Although the top XRD pattern, shown in Figure 2, is characteristic of a fully crystallized material, the Mössbauer spectrum of the same sample S1 treated at 763 K is not quite well resolved suggesting that this sample S1 as well as sample S2 were not fully crystallized, see later.

### 3.2. VM RESULTS

Figure 3 shows the results of the VM measurements for samples S1 and S2, respectively. Notice that for sample S1, the VM value (11.16 GPa) measured at 523 K, is practically the same (11.13 GPa) as the one value measured for the partially crystalline sample. The hardening of S1 starts at 523 K, but below 523 K, the VM values are close to those of the amorphous state, except that at 497 K, which is even lower than that of the as-quenched ribbon. From 296 to 497 K it is possible that stress relief takes place [13]. For sample S2 the VM measurements tend to behave like those of sample S1 in the temperature range from 296 to 407 K. Also notice from Figure 4 that now at 407 K, the VM value is also lower than that of the as-quenched ribbon, i.e., stress relief may take place. In both samples,

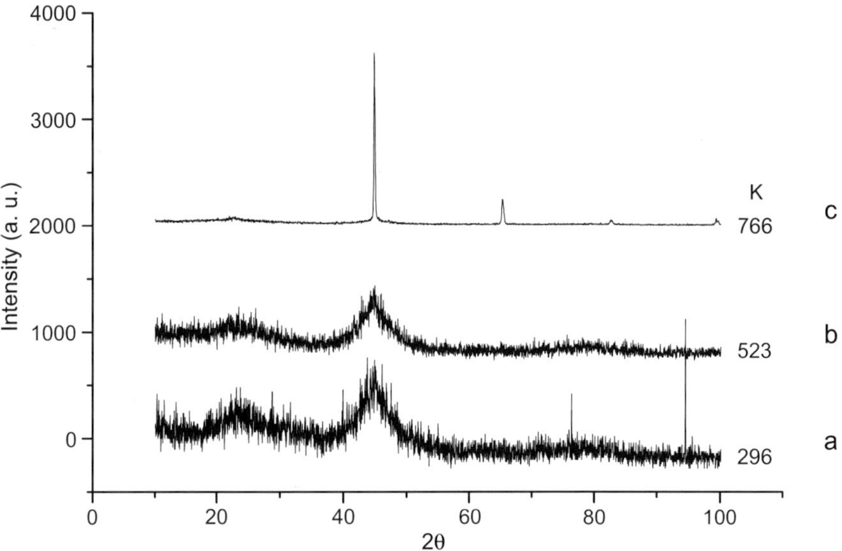


*Figure 2.* XRD patterns of the (a) amorphous sample, (b) sample S1 treated at 523 K, and (c) Crystalline samples S1 and S2. The XRD patterns of crystalline samples S1 ($T_x$ = 763 K) and S2 ($T_x$ = 722 K) were practically the same.

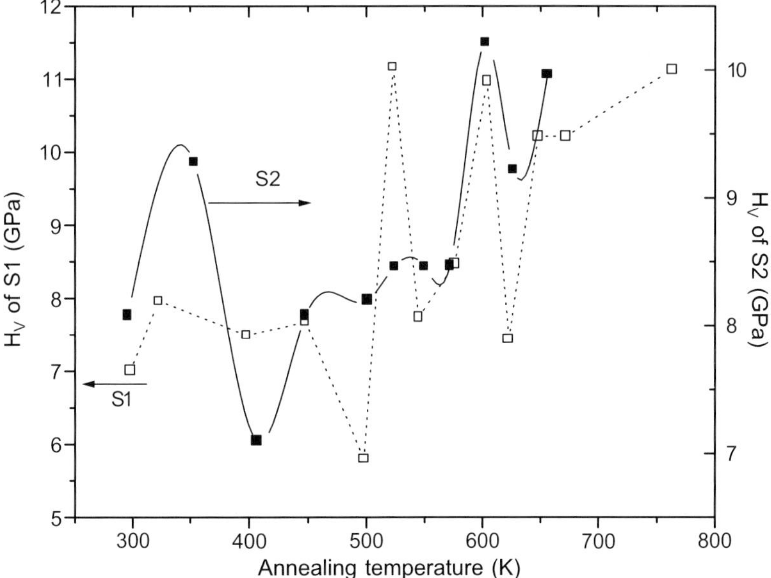


*Figure 3.* VM values of samples S1 and S2 as a function of the annealing temperature. See text.

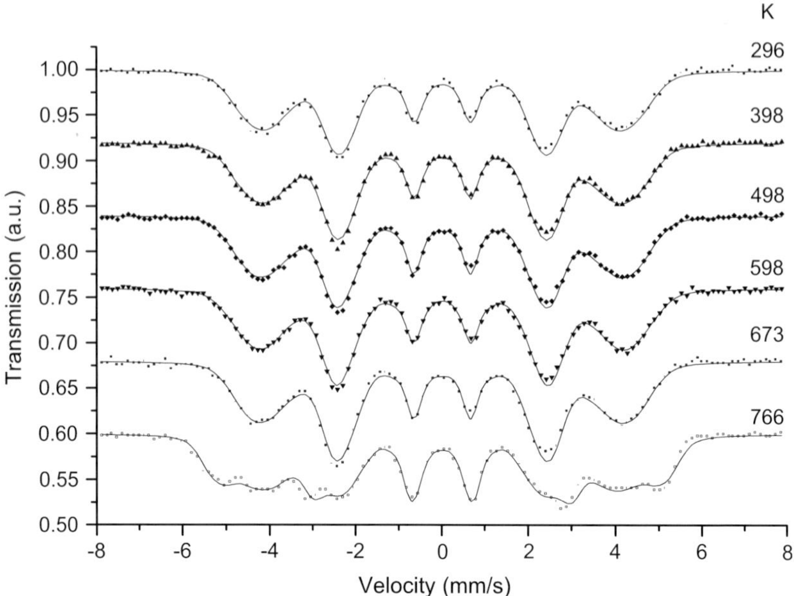


*Figure 4.* Some Mössbauer spectra of the heat treated sample S1.

S1 and S2, the error bars indicate that those lower VM values are statistically significant. On the other hand, in sample S2 the hardening starts at a higher temperature (603 K) than in sample S1 (523 K). Besides, in sample S2 the highest VM value was 10.22 GPa, practically 1 GPa lesser than in sample S1. A significant difference between the partially crystalline samples S1 and S2 was the following. Whereas S1 became fully brittle in its partially crystalline state, S2 remained ductile, slightly less than in the as-quenched sample. This resulting ductility in sample S2 relative to that of sample S1 could be the result of the different cooling rates used. Der-Ray Huang and James C. M. Li [14] report that they completely avoided the brittleness of the amorphous state of $Fe_{78}Si_9B_{13}$ ribbons while treating it with a. c. Joule heating.

The hardening of the amorphous S1 and S2 samples, resulting from present thermal treatments, may be due to other factors than crystalline material, as already shown. Lose of free volume is another source of hardening, which takes place after stress relief [13]. Generally, the melt spinning technique produces stressed metallic ribbons, which may relax by thermal treatments as above. These ribbons are also produced with a characteristic amount of free volume as detected from positron annihilation measurements [6]. Present results, Figure 3, show that this could be the case, since an increase of hardening may represent a compression of the bulk or disappearance of defects. A decrease of hardening may then represent a volumetric expansion of the bulk. A further discussion of this defect aspect is left to section where PALS results are presented.

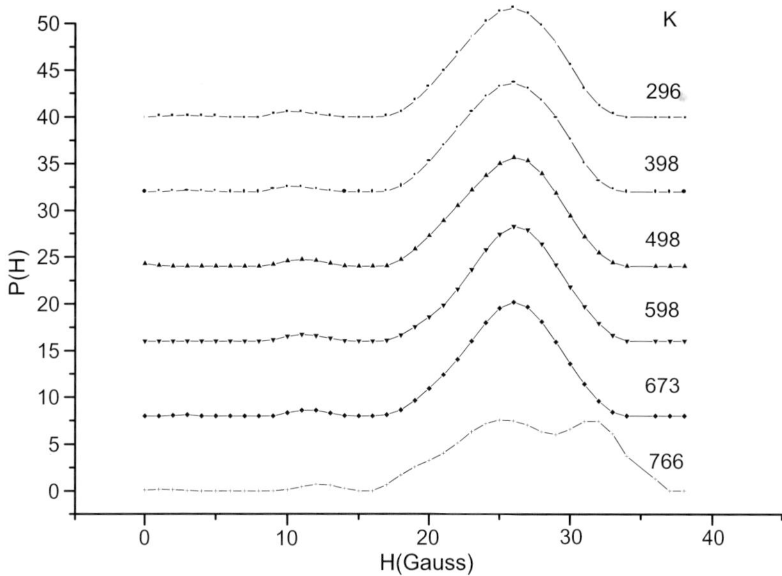


*Figure 5.* The hyperfine magnetic field distributions *P*(*H*) of spectra shown in Figure 4.

### 3.3. MS RESULTS

Figures 4 and 5 show some selected Mössbauer spectra and their corresponding hyperfine magnetic field distributions, P($H$), of ample S1. The Mössbauer spectra consist of broad magnetic sextets with a line intensity ratio close to 3:3:1:1:3:3, characteristic of these metallic amorphous alloys. The Mössbauer line width showed a noticeable annealing temperature dependence; it decreased almost continuously from 296 to 523 K and then it began to increase and decrease in an oscillating form, see Figure 6. This oscillating behavior may be associated to the volumetric compressing and expansion of the sample as stated above; the maximum compaction would be represented by the minimal of the line width values. On the other hand, Figure 7 shows the hyperfine magnetic field ($H$) variation as a function of the annealing temperature. The $H$ value tends to behave with due reserves, as the VM does, compare Figures 3 and 7, i.e., if $H$ increases, VM increases. The magnitude of $H$ may be considered as a measure of strength of the interaction between iron atoms, where the higher the magnitude of $H$, the higher the strength of interaction between iron atoms. In this sense, this could mean that the higher the $H$ value, the lower the mean distance between Fe–Fe atoms, and correspondingly the lower the $H$ value, the higher the mean distance between Fe–Fe atoms. Due to these inter-atomic distance variations, a change in local volume around the iron atoms may also occur. Additional information on this volume variation aspect can be derived from the isomer shift ($\delta$) and the quadruple splitting ($\Delta E_Q$). Figures 8 and 9, show a temperature dependence of these parameters. Both parameters are complementary since they define the

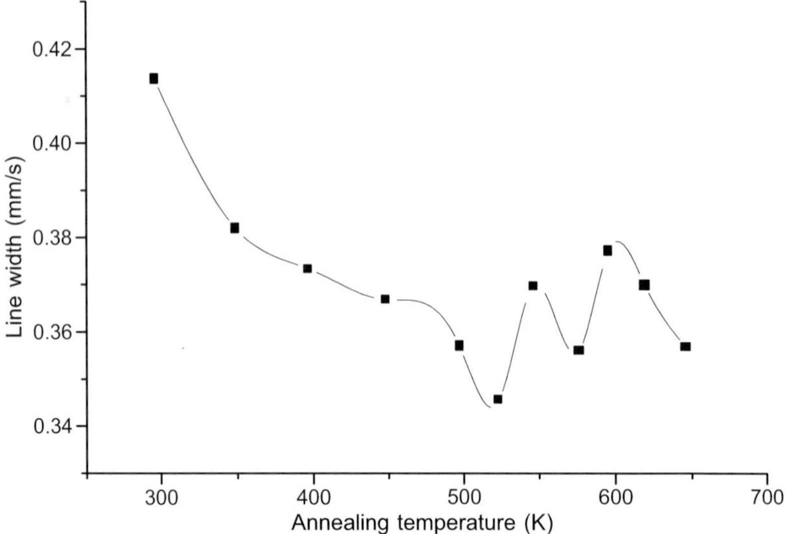


*Figure 6.* Line width of the Mössbauer spectra as a function of the annealing temperature.

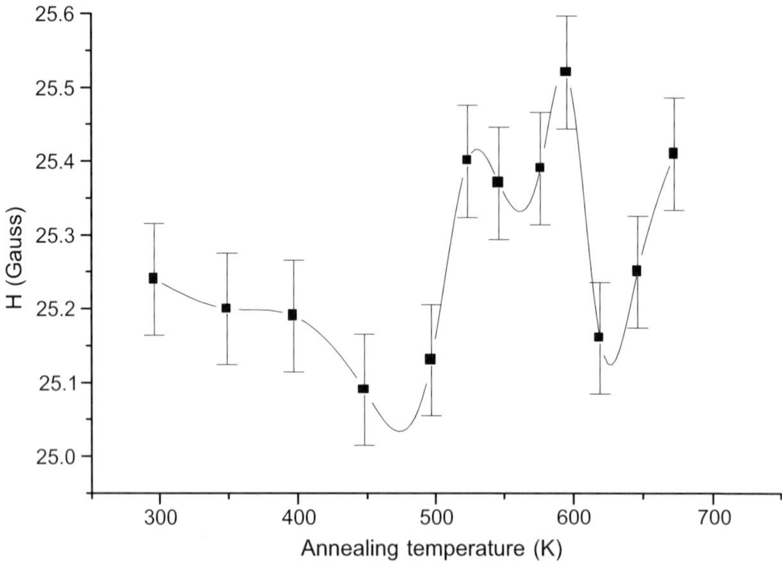


*Figure 7.* The Hyperfine magnetic field, *H*, of sample S1 as a function of the annealing temperature. Notice the resemblance of the *H* temperature dependency with that of the VM one.

electron density at the Fe nucleus and the charge symmetry around the Fe atom, respectively. Generally, the B and Si contents in $Fe_{78}Si_9B_{13}$ tend to increase the value of $\delta$. For instance, whereas in $\alpha$-Fe the $\delta$ is taken to be equal to 0.0 mm/s, in $Fe_{78}Si_9B_{13}$ $\delta = 0.11$ mm/s. That is, the local environment of the Fe atom in

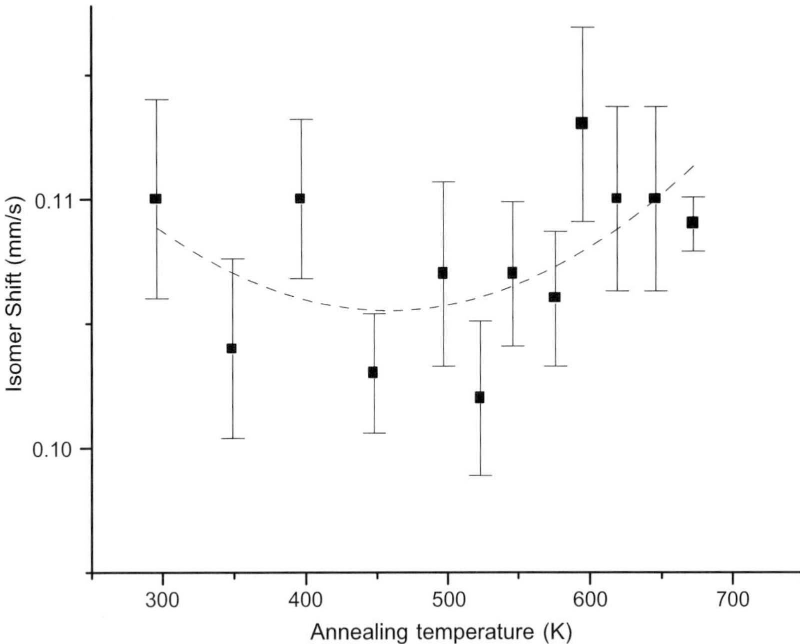


*Figure 8.* The isomer shift, $\delta$, of sample S1 is plotted as a function of the annealing temperature of sample S1. Whereas $\delta$ tends to decrease from 293 to 523 K, it increases from 523 to 673 K. The minimum $\delta$ value occurs at 523 K; this minimum value could mean a volumetric contraction around the iron atom.

$Fe_{78}Si_9B_{13}$ could be between those of $Fe_3B$ and $Fe_2B$ [15]. Present data show that $\delta$ tends to decreases from 296 to 523 K, and increase from 545 to 766 K. The lowest value of $\delta$ occurs at 523 K, which also corresponds to the highest VM and one of the highest B values as shown in Figures 3 and 7, respectively. Also notice from Figure 8 that the $\delta$ values fluctuate between 296 to 523 K stronger than they fluctuate between 545 to 766 K. The faint decrease and oscillating nature of $\delta$ from 296 to 497 K can be associated to the relaxation processes where an increase of the electron density at the iron nucleus is measured. This faint decrease in $\delta$, which is maximum at 523 K, could also be related to a volume contraction around the Fe atom [7] in accord to the high value of $B$ at this temperature. On the other hand, Figure 9 shows the quadruple interaction, $\Delta E_Q$. In this case, the $\Delta E_Q$ values tend to decrease through the whole data points. Notice again that there are differences between the temperature ranges, 296–497 K and 523–766 K, respectively. The fluctuations of $\Delta E_Q$ in the 296–497 K range are greater than in the 523–766 K range. In fact, the tendency of $\Delta E_Q$ below 501 K is to increase faintly and above 497 K to decrease. The fluctuating behavior in the second temperature interval is lower than in the previous one. The net effect of $\Delta E_Q$ suggests that the surrounding of the iron is slightly more symmetric at the end of the experiment than it was in the as-quenched sample.

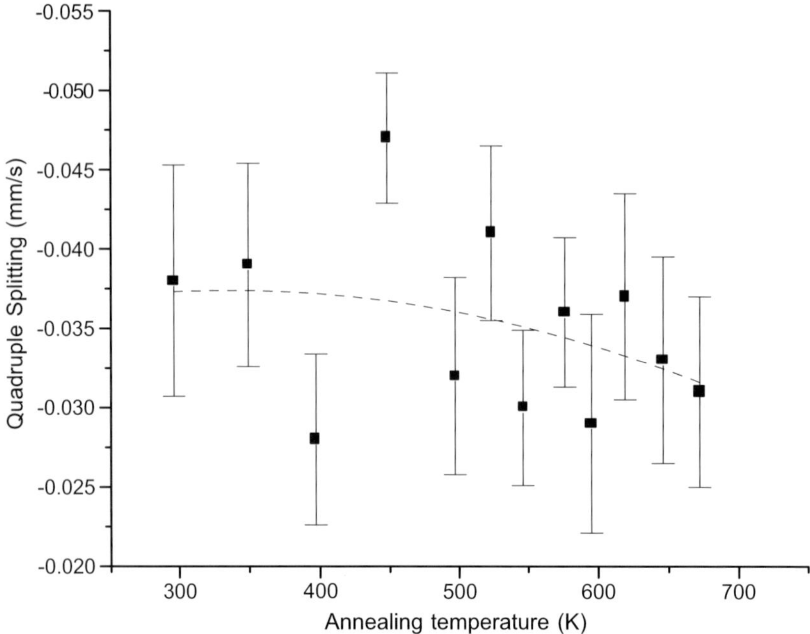


*Figure 9.* The quadruple splitting, $\Delta E_Q$, of sample S1 is plotted as a function of the annealing temperature. Generally, $\Delta E_Q$ decreases from 296 to 673 K. This decreasing tendency of $\Delta E_Q$ suggests that the environment around the iron atom is more symmetric at the end of the experiment than it was for the as-quenched sample.

On the other hand, the widths of the hyperfine magnetic field distributions, $\Gamma_{P(H)}$, were fitted to Gaussian forms and they tend to increase between 296 and 497 K, but decrease between 523 and 766 K. The latter may indicate that the local environment of the iron sites is lesser heterogeneous than in the as-quenched sample. The symmetric forms of the hyperfine magnetic field distributions, Figure 5, also indicate that the local environment of the iron site could be similar throughout the sample. The Mössbauer spectra, recorded in the transmission mode, however, cannot distinguish between the bulk and surface structures of the sample. Thus, from the Mössbauer effect point of view, relaxation and homogenization of sample could only be inferred from 296 to 497 K. At 523 K and above other transformations like creation and annihilation of defects, etc., could take place.

### 3.4. PALS RESULTS

In all cases, three lifetime components fitted the lifetime spectra of sample S3. Figure 10 shows the three lifetime components as a function of the annealing temperature. The average lifetime for $\tau_1$ = 144 ps which is larger than the

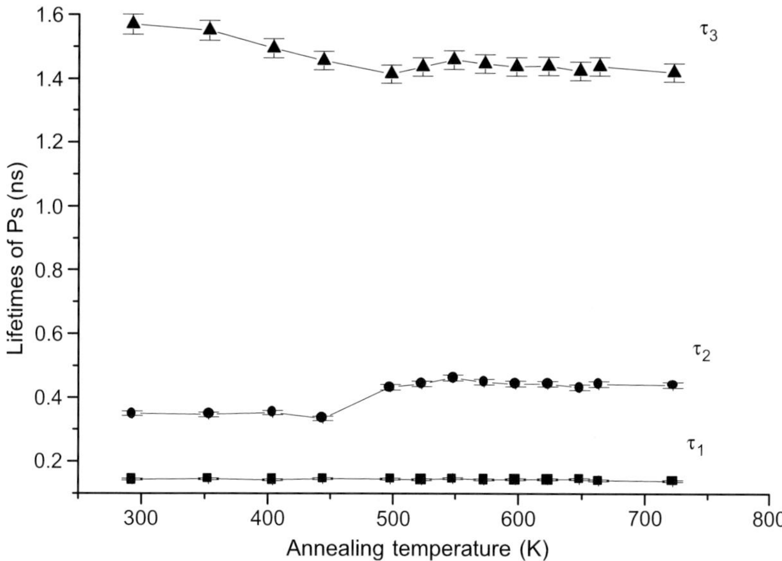


*Figure 10.* The three lifetime components, $\tau_1$, $\tau_2$, $\tau_3$, of sample S3 as a function of the annealing temperature.

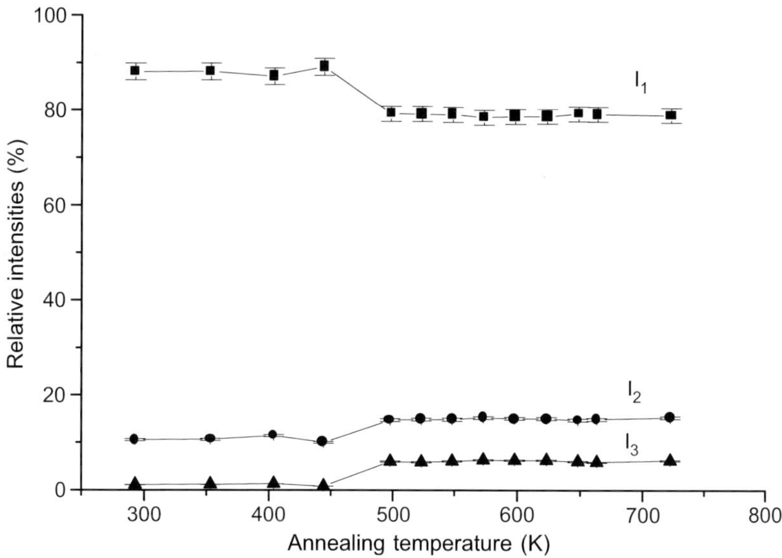


*Figure 11.* The relative intensities, $I_1$, $I_2$, $I_3$, of the three lifetime components of sample S3 as a function of the annealing temperature. See text.

lifetime (113 ps) of well annealed Fe, but shorter than the lifetime (170 ps) for positrons trapped at monovacancies [6]. Even so, this lifetime of 144 ps indicates that in the bulk of the sample S3, where the positrons annihilate, an open volume exists. On the other hand, the relative intensity $I_1$ of $\tau_1$, indicates that between

296 and 448 K it remains practically constant and then it decreases almost 10% at 501 K and finally $I_1$ remains practically constant from 501 to 722 K, Figure 10. This would mean that the net effect of $I_1$ was to decrease about 10% in free volume. $I_1$ also represents the highest contribution (close to 79%) of positrons annihilating in the bulk of $Fe_{78}Si_9B_{13}$ [6]. The second lifetime component, $\tau_2$, as shown in Figure 10, indicates that below and above 501 K behave differently. From 296 to 448 K $\tau_2$ remains practically constant, $\tau_2 \approx 0.35$ ns , but at 501 K it increases markedly with $\tau_2 \approx 0.45$ ns, a value that remains practically constant after 501 K. Again, the number of positrons annihilating with $\tau_2$ between 296 and 448 K remains practically constant and $I_2 \approx 10.5\%$ but $I_2$ increases about a 5% at 501 K and then $I_2$ remains practically constant up to 722 K, Figure 11. The annihilation of positrons with $\tau_2$ may be associated to the annihilation of free positrons near the surface of the ribbons and due to the pick-off process of ortho-positronio (o-Ps). On the other hand, the third lifetime component, $\tau_3$, Figure 10, may describe the annihilation of o-Ps on the surface of the ribbons. In this case, distinct changes also occur between the 296–448 K and between 501–722 K regions. Although $I_3$ is relatively low ($\sim$1%), Figure 11, it remains practically constant in the first temperature interval (296–448 K); $\tau_3$ continuously decreases up to 501 K. From 501 to 722 K $\tau_3$ practically does not change. At 501 K a sudden increase in the intensity of $I_3$ occurs and such an increase is also at the expenses of $I_1$. This means that changes occur on both the inner part and the surface of the sample. Vavassori *et al.* [16] data may support this suggestion. They measured different coercive fields on the surface and in the bulk of the heat-treated amorphous $Fe_{80}B_{20}$ alloy ribbons. According to their magnetic data, magnetic effects of ribbons are stronger on the surface than in the bulk. Whereas the coercive force on the surface increases suddenly after the first heat treatment, in the bulk it increases slowly with the heat treatments.

From the mechanical point of view, and trying to correlate these PALS results of sample S3 with those VM results of sample S1 it could be suggested that in sample S1 the surface changes were more pronounced than in samples S2 and S3. That is, it could be expected that $I_3$ in sample S1 would be higher than it was in sample S3. This speculation will be cleared up in the near future.

### 3.5. CONCLUSIONS

Present experiments have shown that the residual ductility of the partially crystalline $Fe_{78}Si_9B_{13}$ alloy ribbons depend markedly on the previous thermal history. Whereas sample S1 became totally brittle, sample S2 remained ductile in some extent. The spectral line-width and the mean hyperfine magnetic field were more dependent on the thermal treatments than the isomer shift and the quadruple splitting. Additionally, Positron lifetime data also suggested that modifications took place on both the bulk and the surface of the amorphous

$Fe_{78}Si_9B_{13}$ alloy ribbons. These modifications were more pronounced on the surface than in the bulk. Probably the surface modifications are the main cause of the hardening of the alloy in its amorphous state.

## References


1. Devaud-Rzepski J., Quivy A., Harmelin M., Chevalier J.-P. and Calvayrac Y., Electron Microscopy of the initial states of crystalyzation in ferrous glassy alloys: observation of a mode inversion, In: Baró M. D. and Clavaguera N. (eds.), *Current Topics on Non-Crystalline Solids, Proceedings of the First International Workshop on Non-Crystalline Solids*, World Scientific, 1986, pp. 231–236.
2. Zauska A. and Matyja H., Crystallization characteristics of amorphous Fe-Si-B alloys, *J. Mater. Sci.* **18** (1983), 2163–2172.
3. Ramanan V. R. V. and Fish G. E., Crystallization kinetics in Fe-B-Si metallic glasses, *J. Appl. Phys.* **53**(3) (1982), 2273–2275.
4. Rawers J. C., McCune R. A. and Adams A., Crystallization of amorphous $Fe_{78}B_{13}Si_9$, *J. Mater. Sci. Lett.* **7** (1988), 958–960.
5. Lu K. and Wang J. T., Activation energies for crystal nucleation and growth in amorphous alloys, *Mater. Sci. Eng.* **A133** (1991), 500–503.
6. Linderoth S., González J. M., Hidalgo C., Liniers M. and Vicent J. L., Correlation between positron lifetime results and magnetic behaviour upon relaxation and crystallization of an iron-based amorphous alloy, *Solid State Commun.* **65**(12) (1988), 1457–1460.
7. Bang J. Y. and Lee R. Y., Crystallization of the metallic glass $Fe_{78}B_{13}Si_9$, *J. Mater. Sci.* **26** (1991), 4961–4965.
8. Li J.-M., Quan M.-X. and Hu Z.-Q., Spectra investigation on hall-petch relationship in nanocrystalline $Fe_{78}Si_9B_{13}$ alloy, *Appl. Phys. Lett.* **69**(11) (1996), 1559–1561.
9. Chokhsi A. H., Rosen A., Karch J. and Gleiter H., *Scr. Metall.* **23** (1989), 1679.
10. Gleiter H., *Prog. Mater. Sci.* **33** (1989), 223.
11. Weertman J. R., Some unresolved issues concerning mechanical behavior of nanocrystalline metals, *Mat. Sci. Forum* **386–388** (2002), 519–520.
12. Sun X., Cabral-Prieto A., Jose Yacaman M., and Sun W., Investigations on *in situ* nano-crystallization and magnetic properties for amorphous $Fe_{78}Si_9B_{13}$ ribbons, *Mater. Res. Soc. Symp. Proc.* **562** (1999), 301–306.
13. Yavari A. R., Barrue R., Harmelin M. and Perron J. C., *Magn. Magn. Mater.*, **69** (1987), 43.
14. Huang D.-R. and Li J. C. M., High frequency magnetic properties of an amorphous $Fe_{78}Si_9B_{13}$ ribbon improved by a. c. Joule heating, *Mater. Sci. Eng.* **A133** (1991), 209–212.
15. Blum N. A., Moorjani K., Poehler T. O. and Satkiewicsz F. G., Mössbauer investigation of sputtered ferromagnetic amorphous $Fe_xB_{100-x}$ films, *J. Appl. Phys.* **53** (3) (1982), 2074–2076; Chien C. L., Musser D., Gyogy E. M., Sherwood R. C., Chen H. S., Luborsky F. E. and Walter J. L., *Phys. Rev.*, **B20** (1979), 283.
16. Vavassori P., Ronconi F. and Puppin E., Surface crystallization and magnetic properties of amorphous $Fe_{80}B_{20}$ alloy, *J. Appl. Phys.* **82**(12) (1997), 6177–6180.

Hyperfine Interactions (2005) 161:83–92
DOI 10.1007/s10751-005-9170-8 

# Mössbauer and X-Ray Diffraction Investigations of a Series of B-Doped Ferrihydrites

JOHN G. STEVENS*, AIRAT M. KHASANOV and DAVID R. MABE
*Mössbauer Research Group, Department of Chemistry, University of North Carolina at Asheville, One University Heights, Asheville, NC 28804, USA; e-mail: stevens@unca.edu*

**Abstract.** X-ray diffraction and $^{57}$Fe Mössbauer spectroscopy are used to characterize the influence of borate on two-line ferrihydrite's structure and develop likely models for its attachment. Particle sizes were in the 2–4 nm range, and as borate sorption increased, the ferrihydrite particle size decreased. The *d*-spacings of two-line ferrihydrite increased with increased borate adsorption. Isomer shift and quadrupole splitting exhibit slight increasing trends as well. Also, the phase transformation temperature of ferrihydrite to hematite is significantly raised due to borate coating of the surface. We suggest borate is sorbed onto the surface by attachment to the oxygen corners of the iron octahedra that are on the surface of the nanoparticles, placing boron in a tetrahedral molecular geometry.

**Key Words:** borate, ferrihydrite, Mössbauer spectroscopy, nanoparticles, powder X-ray diffraction.

## 1. Introduction

Ferrihydrite is a poorly crystalline iron oxyhydroxide generally given chemical formulas of $Fe_5HO_8{\cdot}4H_2O$ [1] and $5Fe_2O_3{\cdot}9H_2O$ [2, 3], but more recently Eggleton and Fitzpatrick determined the formula to be $Fe_{4.5}(O,OH,H_2O)_{12}$ [4]. Ferrihydrite is categorized by the X-ray diffraction patterns of individual samples: two-line ferrihydrite shows reflections at 0.252 and 0.151 nm, which correspond to the Miller indices of 110 and 300 [3], while six-line ferrihydrite shows additional peaks at 0.223, 0.198, 0.172, and 0.146 nm [3]. Six-line ferrihydrite is a more crystalline version of two-line ferrihydrite; typically, two-line ferrihydrites have particle sizes of 2 nm, while six-line ranges from 2 to 6 nm [2].

The structure of ferrihydrite is believed to be based on that of hematite, with a hexagonal unit cell ($a = 0.508$ nm, $c = 0.94$ nm) [1]. Structurally, ferrihydrite is simple chains of iron octahedra, although tetrahedrally coordinated iron has also been proposed [4, 5]. Pankhurst and Pollard tried to resolve whether or not tetrahedral iron is present in ferrihydrite by studying the Mössbauer spectrum of

* Author for correspondence.

ferrihydrite below its ordering temperature and, using applied fields, determined that there is no appreciable contribution from tetrahedrally coordinated iron [6]. Two-line ferrihydrite consists of iron octahedra that are edge linked into dioctahedral chains of variable length, while six-line ferrihydrite forms from the cross linkage of dioctahedral chains by double corner sharing [3].

Due to its nanoparticle size, ferrihydrite's surface area is very large, giving it high sorptive capabilities. Many heavy metals, important anions, and organic compounds that are of biological significance have been used to study ferrihydrite's sorbent properties. Selenite ($SeO_2^{-3}$) was studied to model possible selenium adsorption routes in animals [7]. One study quantitatively determined how much of a certain anion or cation actually was adsorbed onto the surface of ferrihydrite, including $Ca^{2+}$, $BO_3^-$, and $SO_4^{2-}$ [8]. It is generally recognized that sorption of cation or anion species will raise the phase transformation temperature of ferrihydrite to hematite, which is typically around 570–670 K [3]. In a study on borate sorption to ferrihydrite, it was noted that the phase transformation temperature was raised by hundreds of degrees because of borate coating the surface of the ferrihydrite nanoparticles [9].

Because ferrihydrite mimics the iron nanoparticle core of ferritin, an iron transport protein, it provides a useful model to suggest structural and chemical properties of ferritin's core [1]. Borate is used as a dopant in this study because substantial attention has recently been given to the role of boron in biological systems [10, 11]. While borate doping of ferrihydrite has been discussed before [8, 9], the objective of this study was to carefully construct a picture of how the borate actually attaches to and modifies ferrihydrite's structure using Mössbauer spectroscopy and powder X-ray diffraction.

## 2. Experimental

Two-line ferrihydrite was precipitated from a $FeCl_3$ solution raised to a pH of 8.00 with NaOH. Borate was added in molar ratios $((BO_3/(BO_3 + Fe)) * 100)$ of 0%, 25%, 50%, 75%, and 90% by substituting the necessary amount of $H_3BO_3$ solution for the $FeCl_3$. All samples were washed with distilled water and isopropanol and centrifuged repeatedly until tests for chloride ion with silver nitrate showed no white precipitate. Samples 0A, 1A, 2A, 3A, and 4A were all unheated and were 0%, 25%, 50%, 75%, and 90% borate substituted. Samples 0B, 1B, 2B, 3B, and 4B were all heated to 473 K. Samples heated to 573 K were 0C, 1C, 2C, 3C, and 4C. 0D, 1D, 2D, 3D, and 4D were heated to 673 K. Heat treatments were performed for 4 h at given temperatures in an oven open to air. Samples were weighed before and after heating to investigate percent mass lost after heating.

Powder X-ray diffraction was performed using a Philips X'pert diffractometer with a copper source and only $\alpha 1$ data is reported. A step size of 0.02° (two-theta) and a time per step of 10 s was used. The X-ray diffractograms were fit

*Table I.* Experimental data

| | Mössbauer spectral data | | | | | X-ray diffraction data | | | |
|---|---|---|---|---|---|---|---|---|---|
| | | | | | | 110 Plane | | 300 Plane | |
| | $\delta$ (mm/s) | $\Delta$ (mm/s) | $\Gamma$ (mm/s) | % Area | | *d*-value | Length | *d*-value | Length |
| S | (±0.005) | (±0.005) | (±0.005) | (±5.0) | S | (±0.001) | (±0.10) | (±0.001) | (±0.10) |
| 0A | 0.350 | 0.596 | 0.388 | 79 | 0A | 0.255 | 1.52 | 0.148 | 2.02 |
| | 0.341 | 1.064 | 0.311 | 21 | 0B | 0.251 | 2.1 | 0.147 | 3.84 |
| 0B | 0.344 | 0.601 | 0.478 | 64 | 0C | 0.251 | 3.74 | 0.148 | 2.25 |
| | 0.330 | 1.058 | 0.474 | 36 | 0D | 0.256 | 1.36 | 0.147 | 3.25 |
| 0C | 0.346 | 0.584 | 0.426 | 61 | 0E | 0.258 | 1.36 | 0.147 | 2.83 |
| | 0.337 | 1.043 | 0.436 | 39 | 1A | 0.256 | 1.45 | 0.149 | 1.82 |
| 1A | 0.352 | 0.584 | 0.382 | 70 | 1B | 0.253 | 1.67 | 0.147 | 2.81 |
| | 0.356 | 1.007 | 0.347 | 30 | 1C | 0.251 | 1.97 | 0.149 | 2.05 |
| 1B | 0.349 | 0.636 | 0.428 | 58 | 1D | 0.252 | 4.03 | 0.148 | 2.5 |
| | 0.339 | 1.122 | 0.442 | 42 | 1E | 0.257 | 1.34 | 0.146 | 2.52 |
| 1C | 0.345 | 0.596 | 0.415 | 60 | 2A | 0.259 | 1.43 | 0.149 | 1.39 |
| | 0.336 | 1.057 | 0.468 | 40 | 2B | 0.255 | 1.54 | 0.149 | 1.37 |
| 2A | 0.355 | 0.594 | 0.368 | 68 | 2C | 0.253 | 1.5 | 0.149 | 1.92 |
| | 0.352 | 1.019 | 0.327 | 32 | 2D | 0.254 | 2.75 | 0.148 | 1.49 |
| 2B | 0.349 | 0.656 | 0.457 | 59 | 2E | 0.253 | 3.16 | 0.151 | 2.49 |
| | 0.340 | 1.217 | 0.442 | 41 | 3A | 0.265 | 1.38 | 0.151 | 1.13 |
| 2C | 0.346 | 0.645 | 0.410 | 54 | 3B | 0.259 | 1.4 | 0.149 | 1.29 |
| | 0.337 | 1.151 | 0.464 | 46 | 3C | 0.256 | 1.2 | 0.15 | 1.74 |
| 3A | 0.361 | 0.605 | 0.398 | 50 | 3D | 0.257 | 0.98 | 0.148 | 1.37 |
| | 0.360 | 0.981 | 0.423 | 50 | 3E | 0.254 | 2.33 | 0.148 | 2.29 |
| 3B | 0.364 | 0.691 | 0.429 | 47 | 4A | 0.274 | 1.18 | 0.152 | 1.09 |
| | 0.344 | 1.094 | 0.588 | 53 | 4B | 0.261 | 1.15 | 0.151 | 1.08 |
| 3C | 0.357 | 0.692 | 0.405 | 51 | 4C | 0.262 | 0.92 | 0.151 | 1.42 |
| | 0.354 | 1.203 | 0.469 | 49 | 4D | 0.259 | 0.81 | 0.15 | 0.96 |
| 4A | 0.365 | 0.531 | 0.367 | 40 | 4E | 0.256 | 1.88 | 0.149 | 2.01 |
| | 0.367 | 0.913 | 0.416 | 60 | | | | | |
| 4B | 0.354 | 0.624 | 0.394 | 51 | | | | | |
| | 0.349 | 1.051 | 0.440 | 49 | | | | | |
| 4C | 0.351 | 0.731 | 0.411 | 56 | | | | | |
| | 0.343 | 1.226 | 0.445 | 44 | | | | | |

The first row of Mössbauer data is inner doublet, second is outer doublet. All X-ray data is given in nm. Length was calculated using the Scherrer formula. $S$ = sample name.

with a Lorentzian line shape. *d*-values were obtained using the Bragg equation and particle size using the Scherrer formula. Mössbauer $^{57}$Fe measurements were performed at 297 K using a constant acceleration spectrometer with a $^{57}$Co source in a rhodium matrix that was calibrated using α-Fe. All spectra were fit using a single doublet as well as two doublets. Mössbauer spectral data and X-ray diffraction data are reported in Table I.

## 3. Results

After samples were allowed to dry completely, there was a noticeable color difference. 0A was dark brownish-red, consistent with published color data for two-line ferrihydrite [2], and as borate substitution increased the color became a much lighter brown, also noted by [9]. X-ray patterns of all samples except 0D, 0E, and 1E indicate the main phase present is two-line ferrihydrite because of the broad peaks at 0.255 and 0.148 nm (Figures 1 and 2). Also present is an unknown phase (Figure 1). Diffractograms of samples 0D, 0E, and 1E show the main phase is hematite (Figure 2). Common iron oxide phases and their X-ray diffraction *d*-values and Mössbauer parameters are given in Table II. The phase present in all samples does not match any of the established iron oxides, having X-ray diffraction peaks at 0.234, 0.203, 0.143, and 0.122 nm. X-ray line widths indicate the ferrihydrite particles are about 2 nm, which is characteristic of two-line ferrihydrite [2]. As noticeable from Figure 1, there is a distinct shift of the main ferrihydrite peak at 0.255 nm as well as the peak at 1.48 nm to higher *d*-values as borate concentration increases. For the unheated samples, the 110 plane shifts from 0.255 to 0.274 nm while the 300 plane shifts from 0.148 to 0.152 nm. The average particle size also changes along each plane as borate substitution increases in the unheated samples, with the 110 plane decreasing from 1.52 to 1.18 nm and the 300 plane decreasing from 2.02 to 1.09 nm.

With increasing temperature the particle size increases. 0A had an average particle size of 1.52 nm along the 110 plane, while 0C's average particle size was 3.74 nm. The same trend is noticed in substituted samples. After heating treatments, the trends of increasing *d*-value and decreasing particle size are still noted for both the 110 and 300 planes.

0A's Mössbauer spectrum (Figure 3) gives a paramagnetic doublet with parameters $\delta$ 0.36 mm/s, $\Delta$ 0.69 mm/s, and $\Gamma$ 0.46 mm/s, but when fit with two doublets, has parameters of $\delta$ 0.36 mm/s, $\Delta$ 0.60 mm/s, and $\Gamma$ 0.39 mm/s for the inner doublet, while the outer doublet has $\delta$ 0.34 mm/s, $\Delta$ 1.06 mm/s, and $\Gamma$ 0.31 mm/s. $\Gamma$ values are slightly more than the $\alpha$-Fe calibration material with $\Gamma$ 0.25 mm/s. These parameters are similar to previously published data for two-line ferrihydrite fitted with one [2] and two doublets [12, 13]. During the fitting process, a component was not introduced for the unknown iron oxide phase because using only one and two doublets, chi$^2$ values were very acceptable (0.5 to 1.5) and provided a logical fit to the spectrum. Also, based on X-ray diffraction, the unknown is present in only a few percent of the total material, so if iron is present, it is not dense enough to be seen in a Mössbauer spectrum.

By comparing the spectra from 0A to 4A, the outer doublet shows a slight increase in isomer shift by 0.02 mm/s and the inner doublet has an increase in isomer shift by 0.03 mm/s. Also apparent from comparison is the change in peak area for each doublet as borate substitution increases (Figure 3). As borate substitution increases in unheated samples, the inner doublet's peak area decreases by 40%, while the outer increases by the same amount. There was a

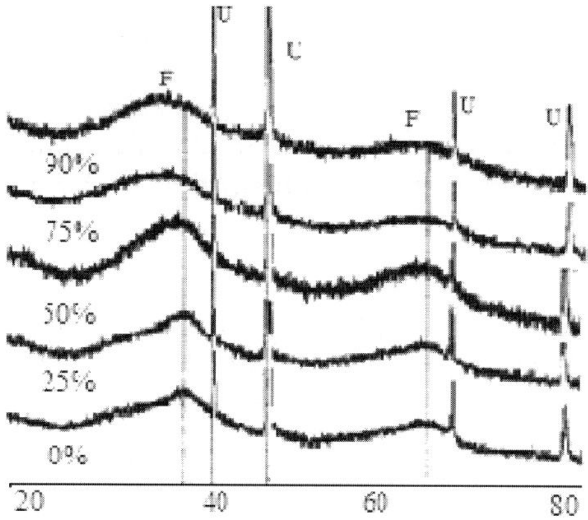


*Figure 1.* X-ray diffractograms of 0A–4A. *F* = Ferrihydrite. *U* = Unidentified iron oxide phase.

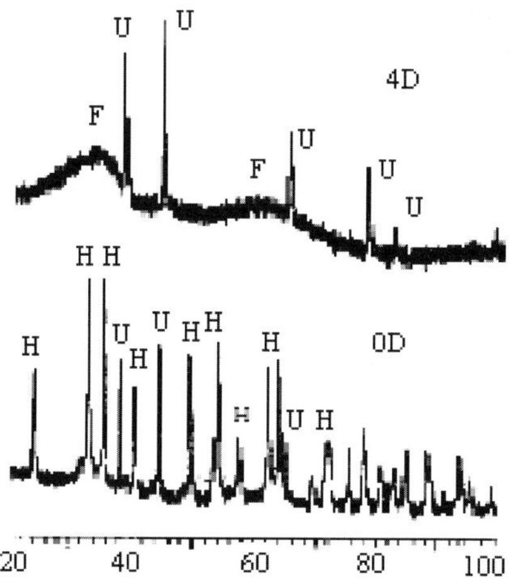


*Figure 2.* X-ray diffractograms of 0D and 4D. *F* = Ferrihydrite, *H* = Hematite, *U* = Unidentified iron oxide phase. Scale is in two-theta degrees.

noticeable increase in Δ from sample 0A (Δ 0.69 mm/s) to 4A (Δ 0.80 mm/s). Also, heated samples showed higher Δ values than non-heated.

After heating all samples to 673 K, only 0D turned partially to hematite (Figure 4). 0D's Mössbauer spectrum consisted of a paramagnetic doublet with parameters $\delta$ 0.33 mm/s, Δ 0.72 mm/s, Γ 0.62 mm/s, and area 16%, which is attributable to ferrihydrite, and a magnetically ordered material with parameters $\delta$ 0.38 mm/s, Δ −0.20 mm/s, Γ 0.43 mm/s, $H$ = 50.9 T, and area 84%, characteristic of hematite. After heating to 773 K, 0E and 1E turned to hematite, while samples 2E, 3E, and 4E stayed two-line ferrihydrite. After heating to 473 K, 0A lost 13% of its weight, while 4A lost 23.5%. Average percent mass lost, assuming hydroxyl groups and water were only groups eliminated

*Table II.* Common iron oxides and their X-ray diffraction and Mössbauer parameters at 297 K

| Mineral name | Compound | XRD spacings in decreasing intensity | | | | | $\delta$ (mm/s) | $\Delta$ (mm/s) | HO (T) |
|---|---|---|---|---|---|---|---|---|---|
| Ferrite | Iron, syn | 2.03 | 0.91 | 1.17 | 1.43 | 1.01 | 0.00 | 0.00 | 33 |
| Magnetite | Magnetite | 2.53 | 1.49 | 1.1 | 1.62 | 2.98 | 0.27 | – | 49 |
| | | | | | | | 0.67 | – | 46 |
| Hematite | Hematite | 2.69 | 1.69 | 2.51 | 1.84 | 1.48 | 0.39 | −0.18 | 52 |
| Goethite | Goethite | 4.18 | 2.69 | 2.45 | 1.72 | 2.19 | 0.35 | 0.54 | 38 |
| Akaganeite | Akaganeite | 7.4 | 3.31 | 1.63 | 2.54 | 1.94 | 0.37 | 0.50 | – |
| Lepidocrocite | Lepidocrocite | 6.26 | 3.29 | 2.47 | 1.94 | 1.73 | 0.38 | 0.27 | – |
| Maghemite | Maghemite | 2.52 | 1.48 | 1.61 | 2.95 | 2.08 | 0.30 | 0.00 | 51 |

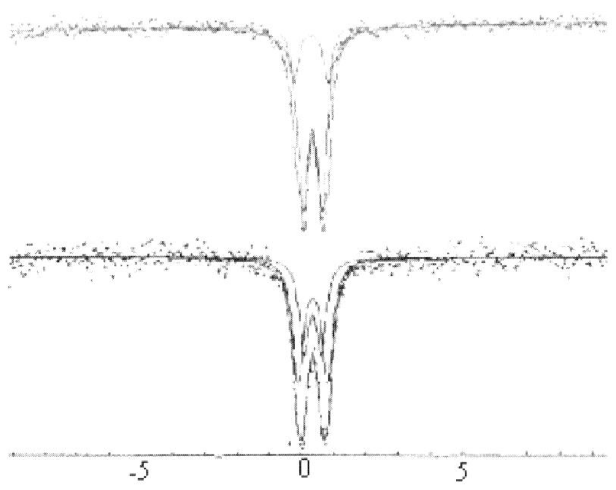


*Figure 3.* 0A and 4A's Mössbauer spectra. Velocity is given in nm/s.

from heating, agrees with chemical formulas given by [1–3]. Samples with a greater percentage of borate lost a greater percentage of weight after heating (Figure 5).

## 4. Discussion

Increased borate adsorption to two-line ferrihydrite increases the *d*-spacing noticeably for the 110 (0.255 nm) plane and increases it slightly for the 300 plane (0.148 nm). Because X-ray diffraction provides an average *d*-spacing for a plane, the actual ferrihydrite iron–oxygen, oxygen–oxygen, or iron–iron distances may not be changing significantly, but rather attachment of borate is causing the average values to be higher. As seen from Table I, lattice spacings seem to be decreasing with increasing temperature, which would be expected because as water and hydroxyl groups are eliminated, the average spacing would decrease along each plane. Particle size decreases with increased borate adsorption,

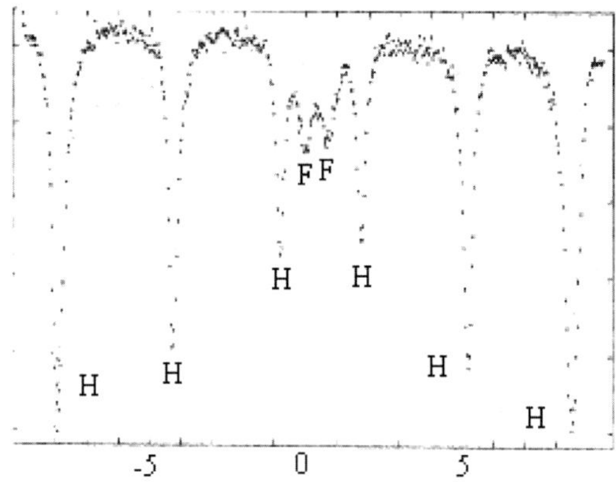


*Figure 4.* Mössbauer spectrum of 0D. *H* = Hematite, *F* = Ferrihydrite. Velocities are mm/s.

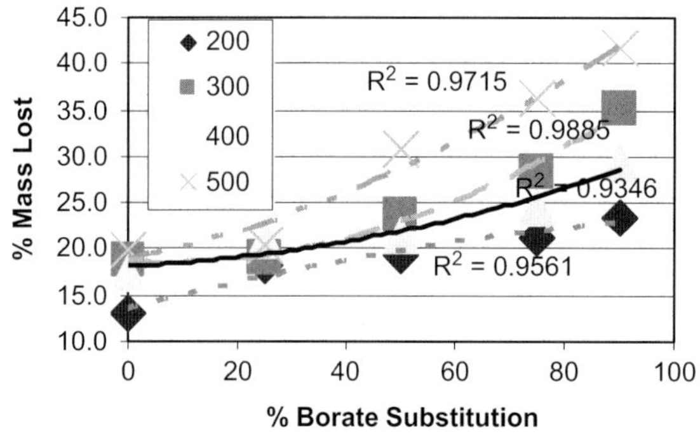


*Figure 5.* Percent mass lost after heating. Legend is in degrees Kelvin.

confirming a previous study indicating that addition of electrolytes to hydrous ferric oxide suspension results in the formation of smaller aggregates [14].

Use of borate as a dopant raises the possibility of the formation of various iron borates (Fe–B–O structure) that contain $Fe^{3+}$ in octahedral sites affected by B in the next nearest environment. An investigation of Mössbauer spectra for the possible structures reveals that none are present in the samples prepared for this investigation. $FeBO_3$ shows a single component with $\delta$ 0.23 mm/s and $\Delta$ 0.19 mm/s [15], both much less than that of the observed values of $\delta$ 0.35 mm/s and $\Delta$ 0.69 mm/s for sample 0A. Warwickite, $Fe_2BO_4$, includes iron in the 3+ and 2+ charge states at 270 K, each of which show two paramagnetic doublets [16]. Vonsenite, $Fe_3BO_5$, includes iron in the 3+, 2.5+, and 2+ charge states, and shows magnetically ordered sextets as well as paramagnetic doublets [16]. Because of the well-defined, easily identifiable multi-component Mössbauer spectra of these iron borates and the required high temperatures ($T > 800$ K) for their synthesis [16, 17], there is no reason to propose their formation in this investigation.

Interpretation of Mössbauer spectra for ferrihydrite varies considerably in the literature. Single doublet fittings produce average values for ferrihydrite and are still necessary as a first step in fitting [18]. Two doublet fittings produce a much better description of the spectrum and have a logical connection to the structural properties of the ferrihydrite [19]. Because of its extremely small particle size, a significant fraction of the iron octahedra of ferrihydrite always represent those of the edges rather than those of the inner structure [3]. 'Edge' iron octahedra are linked to fewer iron octahedra compared to the core iron octahedra, which is why Mössbauer spectra produce parameters closer to the reality of the particle's structure with two doublet fittings [13, 19–21].

If borate is added to the two-line ferrihydrite nanoparticles at oxygen sites on the individual octahedra and increasing the average *d*-values, a slight increase in Mössbauer isomer shift would be expected as borate adsorption increases, which is observed. Single doublet fitting shows a trend towards increasing values, with sample 0A $\Delta$ 0.35 mm/s and 4A $\Delta$ 0.37 mm/s. For two doublet fitting, the inner doublet increases by 0.01 mm/s, while the outer doublet's trend towards increasing values is much more noticeable at 0.03 mm/s. A slight trend towards increasing values for isomer shift is expected and observed because the individual octahedra are not changing, but their immediate chemical environment is. If the actual octahedra were being changed, a greater change would be expected.

Because X-ray diffraction data shows a substantial decrease in particle size from unsubstituted samples to substituted samples, the spectral area for the two doublets in the Mössbauer fittings would be expected to change. Since the inner doublet represents the inner octahedra and the outer doublet represents the edge octahedra, the outer doublet would be expected to increase in area as borate adsorption increases. As particle size decreases, the ratio of edge to inner octahedra becomes greater. If borate molecules are attaching to the available oxygens in the ferrihydrite structure, particle size would be expected to decrease as more borate attaches, because the $B0_3^-$ groups physically restrict where iron octahedra can link together.

After heating below phase transformation temperature to hematite to induce water and hydroxyl group loss, samples with more borate substitution lost a greater mass percentage. Because mass lost is attributable to water and hydroxyl groups, a smaller particle size explains the increased water lost. More surface area provides more room for adsorption and trapping of water molecules.

At ferrihydrite's reported phase transformation temperature of 673 K [3], only the non-substituted sample (0D) turned mostly to hematite. Samples 1D–4D showed no hematite phase present in X-ray diffraction nor Mössbauer data, consistent with the observation that adsorption of ions to ferrihydrite retards its transformation to hematite [3, 9].

If borate is noticeably frustrating the lattice structure of two-line ferrihydrite and decreasing the particle size, the question naturally raised is how it is

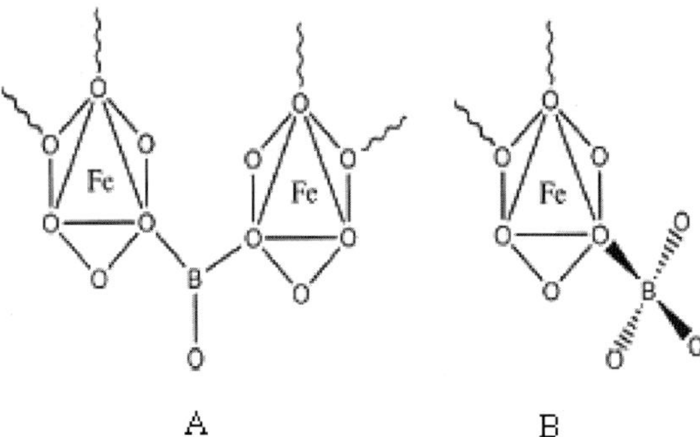


*Figure 6.* Trigonal planar (A) and tetrahedral (B) arrangement of borate groups attached to iron octahedra of ferrihydrite. *Curved lines* represent attachments to other octahedral groups.

incorporated into the structure. Past studies suggest that it attaches to the surface of the particles [9], which is consistent with other experimental data concerning boron's attraction for hydroxyl groups [11]. The attachment to the surface acts as a coating, causing the phase transformation temperature of the ferrihydrite to hematite to increase and result in smaller particle sizes. Two possible boron coordinations exist since boron is stable in trigonal planar and tetrahedral formation (Figure 6). One possibility for borate attachment is that a borate oxygen could be used in the iron octahedra that make up two-line ferrihydrite (Figure 6A). However, this is unlikely because it would be expected that this arrangement could act as a bridge between particles, in effect making particle sizes larger. A more appropriate model is that the central boron of borate is attacked by the free lone electron pairs of outer oxygens of the ferrihydrite octahedra, producing the arrangement in Figure 6B. The slight trend towards higher $\Delta$ values indicates that the actual coordinations of the iron octahedra are not changing, but the chemical environment surrounding them is. The increase in $\delta$ shows that the regular ferrihydrite structure is being frustrated by attachment of borate to the surface, which was also noted by [9].

## References

1. Towe K. M. and Bradley W. F., *J. Coll. Int. Sci.* **24** (1967), 384.
2. Schwertmann U. and Cornell R. M., *Iron Oxides in the Laboratory: Preparation and Characterization*, Weinheim, 2000.
3. Jambor J. and Dutrizac J., *Chem. Rev.* **98** (1998), 2549.
4. Eggleton R. A. and Fitzpatrick R. W., *Clay Miner.* **36** (1988), 111.
5. Cardile C. M., *Clay Miner.* **36** (1988), 537.
6. Pankhurst Q. A. and Pollard R. J., *Clay Miner.* **40** (1992), 268.
7. Parida K. M. *et al.*, *J. Coll. Int. Sci.* **185** (1997), 355.
8. Dzombak D. A. and Morel F. M., *Surface Complexation Modeling: HFO,* New York, 1990.

9. Stevens J. G., Khasanov A. M. and White M. G., *Hyp. Inter.* **151/152** (2003), 283.
10. Barr R. D., Barton S. A. and Schull W. J., *Med. Hyp.* **46** (1996), 286.
11. Nielsen F. H., *Env. Hlth. Persp. Suppl.* **102** (1994), 59.
12. Murad E. and Schwertmann U., *Am. Miner* **65** (1980), 1044.
13. Gotic M. *et al.*, *J. Mat. Sci.* **29** (1994), 2474.
14. Hofmann A. *et al.*, *J. Col. Int. Sci.* **271** (2004), 163.
15. Pollack H. *et al.*, *J. Phys. (Paris) Colloq. C-6* **37** (1976), 585.
16. Mitov I. G. *et al.*, *J. Alloys Cmpd.* **289** (1999), 55.
17. Douvalis A. P. *et al.*, *Hyp. Inter.* **126** (2000), 319.
18. Murad E., *Phys. Chem. Miner.* **23** (1996), 248.
19. van der Kraan A. M., *Phys. Stat. Sol.* **18** (1973), 215.
20. Johnston J. H. and Logan N. E., *J. Chem. Soc. Dalton Trans.* **13** (1979), 13.
21. Murad E. *et al.*, *Clay Miner.* **23** (1988), 161.

Hyperfine Interactions (2005) 161:93–97
DOI 10.1007/s10751-005-9171-7 

# Study of Malayaite and Malayaite Cobalt Pigment

C. PIÑA[1,*], H. ARRIOLA[1] and N. NAVA[2]
[1]*Facultad de Química, Universidad Nacional Autónoma de México, Coyoácan 04510, México, D.F., Mexico; e-mail: cppqic@servidor.unam.mx*
[2]*Programa de Ingeniería Molecular, Instituto Mexicano del Petróleo, Eje Central Lázaro Cárdenas No. 152, Col Atepehuacan, C.P. 07730, México, D.F., Mexico*

**Abstract.** Calcium tin silicate, $CaSnSiO_5$, called Malayaite is synthesized with equimolecular quantities of calcium oxide, silica and stannic oxide followed by a thermic process. In this work, the synthesis of Malayaite and the structure of a Malayaite-based pigment, Sn/Co pink, is investigated by X-ray diffraction and Mossbauer spectroscopy. The results indicate Malayaite and Cassiterite formation, but the ion cobalt incorporated in the Malayaite structure, diminishes the Cassiterite proportion and causes larger asymmetry in the environment of the tin atom.

## 1. Introduction

Malayaite, calcium tin silicate $CaSnSiO_5$, is synthesized in solid state, with stochiometric quantities of calcium oxide, silica and stannic oxide reagents. On the other hand, the effect of different transition metal ion dopants as Cr on the formation of Malayaite has been studied [1–3].

In this work, the synthesis of Malayaite and Malayaite cobalt pigment (pink) is investigated by X-ray diffraction and Mossbauer spectroscopy.

## 2. Experimental

The raw materials used in the preparation of the calcium tin silicate, $CaSnSiO_5$ (Malayaite) and $CaSnSiO_5$–Co(II) (Malayaite cobalt pigment) were $CaCO_3$, $SnO_2$, $SiO_2$ and $CoCl_2{\cdot}6H_2O$ Baker R. A.

The calcium tin silicate was prepared by mixing the starting materials in stochiometric relations in an agate mortar and homogenized in acetone. Afterwards, the sample was calcined in an electric furnace at 1100 °C for 24 h. The synthesis of $CaSnSiO_5$–cobalt pigment was prepared by adding 0.05 mol of $CoCl_2{\cdot}6H_2O$, to the starting materials of Malayaite using the same procedure.

* Author for correspondence.

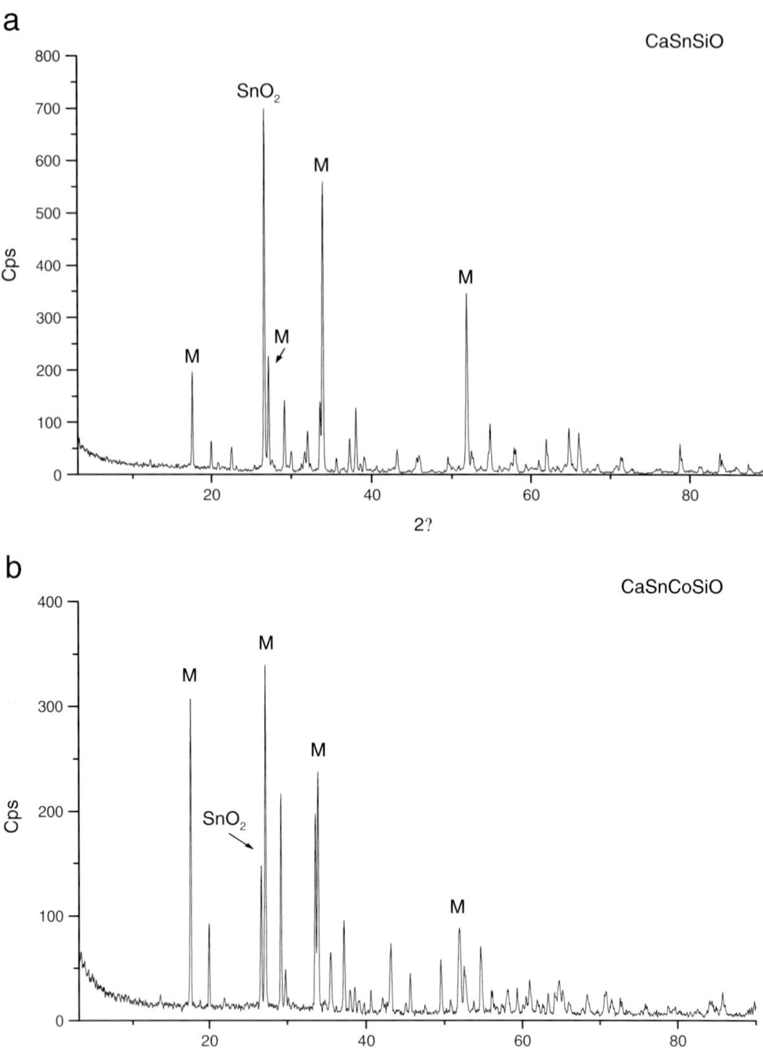


*Figure 1.* (a) XRD pattern of Malayaite (M) showing the peak (110) of Cassiterite ($SnO_2$). (b) XRD pattern of Malayaite cobalt pigment (M) showing the peak (110) of Cassiterite ($SnO_2$).

## 2.1. X-RAY POWDER DIFFRACTION

X-ray diffraction patterns of fired samples were obtained in a Siemmens D5000 diffractometer using Cu, K$\alpha$ radiation ($\lambda$ = 1.5406 Å) and a Ni filter in a 5° (2$\theta$) 85° range.

## 2.2. $^{119}$SN MOSSBAUER SPECTROSCOPY

Mossbauer spectra were carried out using a constant acceleration Mossbauer Austin Scientific Associates S-600 spectrometer and a 10 mCi $^{119}$Sn source,

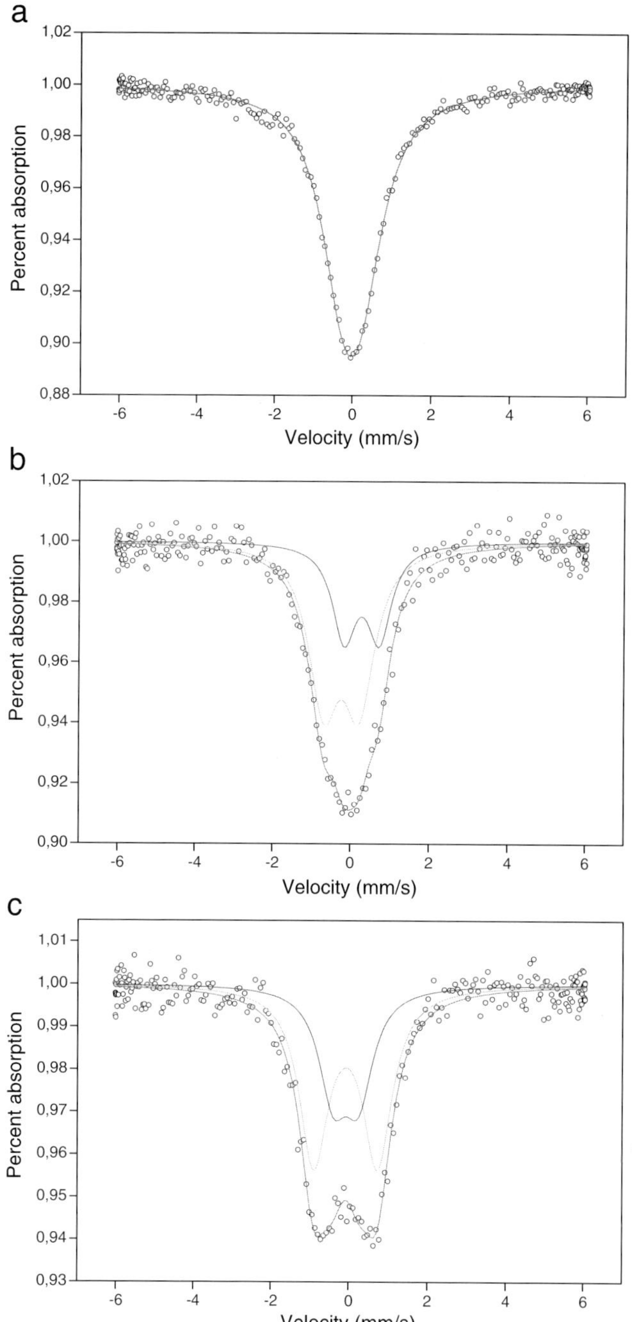


*Figure 2.* (a) Mossbauer spectrum of $SnO_2$. (b) Mossbauer spectrum of Malayaite and Cassiterite. (c) Mossbauer spectrum of Malayaite cobalt pigment and Cassiterite.

*Table I.* Isomeric shift and quadrupole splitting of $SnO_2$, Malayaite–$SnO_2$ and Malayaite Cobalt pigment–$SnO_2$

| | |
|---|---|
| $SnO_2$ | $\delta = -0.019$ mm/s<br>$\delta = 0.510$ mm/s |
| Malayaite and $SnO_2$ | $\delta = -0.225$ mm/s<br>$\delta = 0.902$ mm/s<br>$\delta = 0.285$ mm/s<br>$\delta = 0.891$ mm/s |
| Malayaite Co pigment and $SnO_2$ | $\delta = -0.079$ mm/s<br>$\delta = 1.623$ mm/s<br>$\delta = -0.075$ mm/s<br>$\delta = 0.649$ mm/s |

calibration was carried out with a $\beta$-Sn (white) foil, its $\delta = 2.56$ mm/s with respect to $SnO_2$, all the spectra were taken at room temperature and the velocity range was from $-6$ to $+6$ mm/s. The Mossbauer spectra were fitted with a Lorentzian line least squares algorithm.

## 3. Results and discussion

X-ray diffraction patterns (Figure 1a and b) Malayaite and stannic oxide (Cassiterite) are observed in both samples. The peak with Miller index (110) corresponding to a $SnO_2$ in the Malayaite synthesis is more intense than in the Malayaite Cobalt pigment. The effect of doping Malayaite with Co(II) shows that the resulting pigment consists of a mixture of Malayaite, a 40% at least of Cassiterite and a change in colour from white to pink. On the other hand the effect of Co(II) on the formation of Malayaite Cobalt pigment, shows that at 0.05 mol concentration, the formation of Malayaite is more favored and the proportion of Cassiterite is reduced (see X-ray diffraction patterns).

The structure of Malayaite consists of infinite chains of vertex-sharing $SnO_6$ octahedral linked by isolated $SiO_4$ tetrahedral, structures of which are appropriate for the incorporation of metal ions coloured pigments [1, 2, 4] as it can be observed in the results obtained with the cobalt ion.

In Figure 2a, b, and c, the Mossbauer spectra for $SnO_2$, M/$SnO_2$, and Co/M/$SnO_2$ respectively are shown. In Table I, the Mossbauer parameters for the different samples are shown, the isomeric shifts are within the tin(IV) range ($-0.5$ to 2.7 mm/s), and no absorption peak was observed in the tin(II) range (2.7 to 5.0 mm/s), which means that there was not any reduction of tin(IV) to tin(II) during the firing process, the presence of even small amounts of tin(II) would have been detected. Figure 2b and c confirm the results of XRD, which show two doublets in each case, for the two phases, Malayaite and Cassiterite.

Quadrupolar splitting for Malayaite pigment, Calcium tin silicate, $CaSnSiO_5$ and for Malayaite Cobalt pigment $CaSnCoSiO_5$ are larger than for $SnO_2$, which shows more asymmetry in the surroundings of the tin atom. As it can be seen the Co ion incorporation to Malayaite increases even more the asymmetry of the compound.

## 4. Conclusions

X-ray diffraction results show that, by addition of a cobalt divalent ion to the starting materials of Malayaite a cobalt pigment is formed, which consists of mixtures of both Malayaite and small quantities of Cassiterite. That means that at 0.05 mol concentration the proportion of Malayaite is increased whereas the stannic oxide decreases. $^{119}Sn$ Mossbauer spectra results indicate the incorporation of cobalt ions in the lattice of Malayaite, which tends to make the tin atom environment more asymmetric.

## References

1. Escribano López P., Guillem Monzones C., and Alarcón Navarro J., $CrSnO_2$–CaO–$SiO_2$ Based Ceramic Pigments, *Ceramic Bulletin* (1984).
2. Bauer, W. H. and Khan A. A., *Acta Cryst. B* **27** (1971), 2133.
3. Takenouchi, *Mineral Deposita* **6** (1971), 335–347.
4. Sanghani D. V., Abrams G. R. and Smith, P. J., *International Tin Research*, Fraser Road, Greenford, Middlesex, UB67AQ, 1981, pp. 311–315.

Hyperfine Interactions (2005) 161:99–111
DOI 10.1007/s10751-005-9172-6


# Study on Stereochemical Activity of Lone Pair Electrons in Sulfur and Halogen Coordinated Antimony(III) Complexes by $^{121}$Sb Mössbauer Spectroscopy

RYUHICHI OHYAMA†, MASASHI TAKAHASHI and MASUO TAKEDA*
*Department of Chemistry, Faculty of Science, Toho University, Miyama, Funabashi, Chiba 274-8510, Japan; e-mail: takeda@chem.sci.toho-u.ac.jp*

*Dedicated to Prof. Philipp Gütlich's 70th birthday.*

**Abstract.** We have measured $^{121}$Sb Mössbauer spectra at 20 K for 52 compounds of antimony(III). An Sb(III) atom with the electron configuration [Kr] $4d^{10}5s^2$ has a lone pair electrons. The stereochemical property of the lone pair has been found to depend very much on the kinds of atoms surrounding the antimony atom and the configurations of the coordinating atoms.

**Key Words:** coordination configuration, lone pair electron, $^{121}$Sb Mössbauer spectroscopy, Sb(III) complexes.

## 1. Introduction

An Sb(III) atom has an electron configuration [Kr] $4d^{10}5s^2$ identical to Sn(II) and Te(IV) atoms. Their compounds have often non-symmetrical coordination polyhedra due to the presence of the stereochemically active lone pair electrons. The Mössbauer spectroscopy of $^{119}$Sn [1, 2], $^{121}$Sb[1, 3–5] and $^{125}$Te [6] has been proved to be very useful to investigate the orbital character of the lone pair.

The $^{125}$Te quadrupole splitting was found to be larger in $TeO_4E$ (Ψ-trigonal bipyramid, lone paired electrons, $E$ is in equatorial position) than in $TeO_6$E (Ψ-monocapped octahedron, $SbO_3^{short}O_3^{long}E$, $E$ is accommodated among the three O atoms at longer distances and along the $C_3$ axis) [6].

We have found that both of the values of the isomer shifts and quadrupole coupling constants of $^{121}$Sb Mössbauer spectra at 20 K, whose decrease means the decrease in the stereochemical activities of the lone pair, decrease in the order:$SbO_3E$ configuration (Ψ-tetrahedron) in $Sb_2O_3$, where $E$ denotes lone pair ≈

* Author for correspondence.

† Present address: Tama Chemicals Co., Ltd. 3-22-9, Shiohama, Kawasaki-ku, Kawasaki City, Kanagawa 210-0826, Japan.

*Table I.* $^{121}$Sb Mössbauer parameters for antimony(III) compounds at 20 K

| | Compound | $\delta$ mm s$^{-1}$ | $e^2qQ$ mm s$^{-1}$ | $\eta$ | $\Gamma_{Exp}$ mm s$^{-1}$ | Reference |
|---|---|---|---|---|---|---|
| 1 | $[Sb(S_2CNMe_2)_2Br]$ | −5.80 | 10.00 | 0.40 | 2.5 | own |
| 2 | $[Sb(S_2CNEt_2)_2Br]$ | −6.30 | 9.30 | 0.50 | 2.8 | own |
| 3 | $[Sb(S_2CNEt_2)_2I]$ | −6.20 | 9.60 | 0.50 | 2.6 | own |
| 4 | $[Sb(S_2CNBu^i{}_2)_2I]$ | −6.10 | 9.20 | 0.50 | 2.6 | own |
| 5 | $[Sb(S_2CN\text{-}Pyr)_2Cl]$ | −6.00 | 11.20 | 0.70 | 2.9 | own |
| 6 | $[Sb(S_2CN\text{-}Pyr)_2Br]$ | −5.50 | 10.20 | 0.50 | 2.4 | own |
| 7 | $[Sb(S_2CN\text{-}Pyr)_2I]$ | −5.90 | 9.70 | 0.50 | 2.5 | own |
| 8 | $[Sb(S_2COMe)_3]$ | −6.37 | 9.90 | 0.31 | 2.36 | own |
| 9 | $[Sb(S_2COEt)_3]$ | −6.23 | 9.20 | 0.12 | 2.31 | own |
| 10 | $[Sb(S_2COPr^i)_3]$ | −5.44 | 9.30 | 0.29 | 2.39 | own |
| 11 | $[Sb(S_2COBu^i)_3]$ | −6.19 | 9.30 | 0.60 | 2.86 | own |
| 12 | $[Sb(S_2COBu^s)_3]$ | −6.32 | 10.40 | 0.74 | 2.63 | own |
| 13 | $[Sb(S_2COEt)_2Br]$ | −6.43 | 9.20 | 0.72 | 2.68 | own |
| 14 | $[Sb(S_2COEt)_2I]$ | −6.12 | 9.20 | 0.61 | 2.49 | own |
| 15 | $[Sb\{S_2P(OMe)_2\}_3]$ | −7.06 | 8.10 | 0.10 | 2.47 | own |
| 16 | $[Sb\{S_2P(OEt)_2\}_3]$ | −7.15 | 8.00 | 0.26 | 2.28 | own |
| 17 | $[Sb\{S_2P(OPr^i)_2\}_3]$ | −7.10 | 8.50 | 0.36 | 2.28 | own |
| 18 | $[Sb\{S_2P(OBu^s)_2\}_3]$ | −7.09 | 7.90 | 0.58 | 2.40 | own |
| 19 | $[Sb\{S_2P(OBu^t)_2\}_3]$ | −7.66 | 7.20 | 0.82 | 2.36 | own |
| 20 | $[Sb(S_2PEt_2)_3]$ | −6.74 | 8.00 | 0.71 | 2.36 | own |
| 21 | $[Sb(S_2PPh_2)_3]$ | −6.00 | 9.10 | 0.29 | 2.28 | own |
| 22 | $[Sb(SBu^t)_3]$ | −2.71 | 12.90 | 0.46 | 2.67 | own |
| 23 | $[Sb(SPent^t)_3]$ | −3.97 | 15.20 | 0.06 | 2.81 | own |
| 24 | $[Sb(SPh)_3]$ | −3.51 | 12.70 | 0.09 | 2.66 | own |
| 25 | $K_3[SbCl_6]$ | −9.99 | 5.10 | 0 | 2.78 | own |
| 26 | $Rb_3[SbCl_6]$ | −10.01 | 4.60 | 0 | 2.82 | own |
| 27 | $Cs_3[SbCl_6]$ | −9.70 | 4.50 | 0 | 2.67 | own |
| 28 | $[Co(NH_3)_6][SbCl_6]$ | −10.78 | 0.40 | 0 | 2.67 | own |
| 29 | $[Sb(S_2CNMe_2)_3]$ | −6.03 | 6.20 | 0.03 | 2.88 | own |
| 30 | $[Sb(S_2CNEt_2)_3]$ | −6.78 | 8.00 | 0.38 | 2.11 | 3 |
| 31 | $[Sb(S_2CNPr^n_2)_3]$ | −6.01 | 8.10 | 0.29 | 2.16 | 3 |
| 32 | $[Sb(S_2CNBu^n_2)_3]$ | −6.34 | 7.20 | 0.41 | 2.31 | 3 |
| 33 | $[Sb(S_2CNBu^i_2)_3]$ | −6.49 | 9.50 | 0.16 | 2.26 | 3 |
| 34 | $[Sb(S_2CNBz_2)_3]$ | −6.93 | 7.00 | 0.47 | 2.33 | 3 |
| 35 | $[Sb(S_2C\text{-}Pyr)_3]$ | −6.32 | 5.90 | 0.06 | 2.52 | 3 |
| 36 | $[Sb(S_2C\text{-}Pip)_3]$ | −6.03 | 8.10 | 0.43 | 2.36 | 3 |
| 37 | $[Sb[Hedta)]\cdot 2H_2O$ | −7.18 | 11.50 | 0.00 | 2.29 | 3 |
| 38 | $Li[Sb(edta)]\cdot 2H_2O$ | −7.27 | 12.70 | 0.18 | 2.83 | 3 |
| 39 | $Na[Sb(edta)\cdot 2H_2O$ | −7.34 | 12.90 | 0.10 | 2.15 | 3 |
| 40 | $K[Sb(edta)]\cdot 2H_2O$ | −7.99 | 10.90 | 0.17 | 2.26 | 3 |
| 41 | $Rb[Sb(edta)]\cdot H_2O$ | −6.82 | 11.10 | 0.02 | 2.35 | 3 |
| 42 | $Cs[Sb(edta)]\cdot H_2O$ | −6.42 | 12.50 | 0.25 | 2.50 | 3 |
| 43 | $NH_4[Sb(edta)]\cdot 0.5H_2O$ | −6.96 | 10.90 | 0.28 | 2.18 | 3 |
| 44 | $K_3[Sb(ox)_3]\cdot 4H_2O$ | −6.60 | 15.10 | 0.07 | 2.32 | 3 |
| 45 | $Rb_3[Sb(ox)_3]\cdot 1.5H_2O$ | −6.49 | 14.10 | 0.20 | 2.58 | 3 |
| 46 | $Cs_3[Sb(ox)_3]\cdot 2H_2O$ | −7.14 | 14.30 | 0.23 | 3.21 | 3 |
| 47 | $K_2[Sb_2(tart)_2]\cdot 4H_2O$ | −3.41 | 19.00 | 0.52 | 2.72 | 3 |

*Table I.* (Continued)

| | Compound | $\delta$ mm $s^{-1}$ | $e^2qQ$ mm $s^{-1}$ | $\eta$ | $\Gamma_{Exp}$ mm $s^{-1}$ | Reference |
|---|---|---|---|---|---|---|
| 48 | $Sb_2O_3$ | −3.33 | 18.10 | 0.31 | 2.40 | 3 |
| 49 | $Sb_2S_3$ | −6.27 | 11.10 | 0.47 | 2.77 | 3 |
| 50 | $AgSbS_2$ | −5.16 | 9.60 | 0.28 | 2.38 | 3 |
| 51 | $KSbS_2$ | −4.18 | 13.00 | 0.86 | 2.40 | 3 |
| 52 | $FeSb_2S_4$ | −5.39 | 11.10 | 0.80 | 3.06 | 3 |

$SbO_4E$ ($\Psi$-trigonal bipyramid, $E$ is in equatorial position) in $[Sb_2(tart)_2]^{2-}$ > $SbO_6E$ ($\Psi$-pentagonal bipyramid, $E$ occupies an axial position) in $[Sb(ox)_3]^{3-}$ ≥ $SbN_2O_4E$ ($\Psi$-pentagonal bipyramid; $E$ located in equatorial plain) in [Sb(Hedta)] and $[Sb(edta)]^-$ [3].

We have also found that for sulfur coordinated Sb(III) atoms, both parameters decrease in the order: $SbS_4E$ ($\Psi$-trigonal bipyramid, $E$ is in equatorial position) in $KSbS_2$ > $SbS_3^{short}S_3^{long}E$ ($\Psi$-monocapped octahedron, $E$ is along $C_3$ axis) in $[Sb(S_2CNR_2)_3]$ (R = Et, $Pr^n$, $Bu^t$, $Bu^i$, Bz), $[Sb(S_2C\text{-}Pyr)_3]$ and $[Sb(S_2C\text{-}Pip)_3]$ [3]. Furthermore we measured $^{121}Sb$ Mössbauer spectra for crown ether adducts of antimony(III) trihalides $SbX_3$(crown) and was found a good relationship between Mössbauer parameters and Sb-O bond lengths. The lone pair has more *p* orbital character as the bond length becomes larger [4, 5].

Though the $^{121}Sb$ Mössbauer parameters of the following compounds have been reported from some laboratories, the isomer shifts must be compared with the values obtained under the same experimental conditions: $Sb_2O_3$, **48** [7, 8]; $[Sb(S_2CNEt_2)_3]$, **30** [9]; $[Sb(S_2CNBu^n_2)_3]$, **32** [9]; $[Sb(S_2CNEt_2)_2Br]$, **2** [9]; $[Sb(S_2CNEt_2)_2I]$, **3** [9]; $Sb_2S_3$, **49** [8]; $K_3[SbCl_6]$, **25** [10]; $Cs_3[SbCl_6]$, **27** [10] and $[Co(NH_3)_6][SbCl_6]$,**28** [10, 11].

So we measured the Mössbauer spectra of 29 compounds which makes the total number of compounds which were measured in our laboratory, 52 to be compared concerning the Mössbauer parameters. We focused on the stereochemical activity of the lone pair in Sb(III) complexes coordinated by sulfur/halogen atoms as well as oxygen/nitrogen atoms.

## 2. Experimental

### 2.1. PREPARATION

We have synthesized $[Sb(SR)_3]$ (R = Ph, $Bu^t$, $Pent^t$), $[Sb(S_2PR_2)_3]$ (R = Et, Ph), $[Sb(S_2CNMe_2)_3]$, $[Sb(S_2COR)_3]$ (R = Me, Et, $Pr^i$, $Bu^i$, $Bu^s$), $[Sb\{S_2P(OR)_2\}_3]$ (R = Me, Et, $Pr^i$, $Bu^s$, $Bu^i$), $[Sb(S_2CN\text{-}Pyr)_2X]$ (X = Cl, Br, I), $[Sb(S_2CNR_2)_2X]$ (R = Me, Et, $Bu^i$; X = Br, I), $[Sb(S_2COEt)_2X]$ (X = Br, I), and $M_3[SbCl_6]$ (M = K, Rb, Cs, $[Co(NH_3)_6]$).

The complex **1** (every complex is denoted by figure as in Table I) was made according to a method similar to that for **2** and **6** [12]. The complexes **3**, **4** and **7** were prepared by the method in Ref. [13]. The complex **5** was prepared as follows. 0.12 ml (2 mmol) of carbon disulfide and 0.20 ml (2 mmol) of pyrrolidine were added to 150 ml carbon tetrachloride solution of antimony(III) chloride of 0.23 g (1 mmol) keeping at −80°C using dry ice-acetone bath. White precipitates obtained after cooling were filtrated and dissolved into 30 ml of chloroform. After separating the residue the filtrate was mixed with 100 ml of carbon tetrachloride to obtain the complex **5** as precipitate. After filtration they were washed by water. White crystals of **5** were obtained by recrystallization from 200 ml of mixed solution of carbon tetrachloride–chloroform (1:1) and dried at reduced pressure. Yield, 0.22 g (49%), m. p., 235–237°C.

The complexes **8**, **10**, **11** and **12** were synthesized according to Ref. [14]. Since the yield of 11.5% for the complex **9** according to the known method [14] was so small, we tried a new synthetic method to make it. 9.60 g (60 mmol) of potassium xanthate was added to 200 ml of benzene solution of 4.56 g (20 mmol) of antimony(III) chloride and stirred for 2 h at room temperature. The solid formed was filtered off. Vacuum concentration of the yellowish filtrate produced sticky orange solution which produced orange-colored yellow crystals after leaving stand at room temperature. They were dried in vacuo. Yield, 4.98 g (51%), m. p., 85–86°C. The complexes **13** and **14** were made according to Ref. [15]. The complexes **15**–**21** were produced as the same method in Ref. [16]. The complex **24** which is very easily oxidized was made under nitrogen atmosphere by the method in Ref. [17], and the complexes **22** and **23** ($Pent^t$ = tert-$C_5H_{11}$, pentyl) were produced as the same method in Ref. [17]. The compounds **25**–**27** which are very sensitive to moisture were produced by the method in Ref. [18]. The compound **28** was prepared according to the method in Ref. [19]. The complex **29** was synthesized by a method similar to that in Ref. [20].

The compounds obtained were all characterized by elemental analyses of C, H, N and Sb, and IR spectroscopy.

### 2.2. MÖSSBAUER MEASUREMENT

$^{121}$Sb Mössbauer spectra were measured using an Austin Science S-600 Mössbauer spectrometer using a $Ca^{121m}SnO_3$ source (16 MBq). Both the sample containing 15 mg Sb cm$^{-2}$ and the source were kept at 20 K in a cryostat equipped with a closed cycle refrigerator. The Doppler velocity was measured with an Austine Science LA-9 laser interferometer and calibrated by recording the $^{57}$Fe Mössbauer spectrum of an α-iron foil at 20 K. Normally it took several days to get a Mössbauer spectrum. The $^{121}$Sb Mössbauer spectra were computer-fitted to quadrupole-split twelve lines using a transmission-integral method [21] on a personal computer. The values of the isomer shift ($\delta$) are given relative to InSb at 20 K.

## 3. Results and discussion

Figure 1 demonstrates the Mössbauer spectra of the compounds with different coordination configurations which are shown in the figure. Figure 2 exhibits the

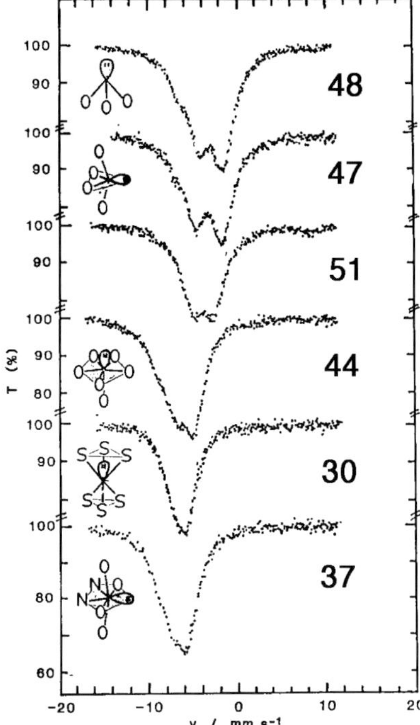


*Figure 1.* $^{121}$Sb Mössbauer spectra of antimony(III) compounds at 20 K.

gradual decrease in the isomer shifts from the top to the bottom; Ψ-tetrahedral **24**, Ψ-pentagonal bipyramidal **21**, slightly distorted octahedral **26** and regular octahedral **28**. Figure 3 demonstrates the gradual increase in the isomer shifts from the top to the bottom in type $SbS_3^{short}S_3^{long}$-S′E for three $[Sb(S_2CNR_2)_3]$ and in type $SbS_3^{short}S_3^{long}E$ for $AgSbS_2$.

Table I summarizes the Mössbauer parameters such as isomer shift ($\delta$), electric quadrupole coupling constant ($e^2qQ$), asymmetry parameter ($\eta$) and experimental half linewidth ($\Gamma_{exp}$) obtained. The $e^2qQ$ value of 12.70 mm $s^{-1}$ obtained by this Mössbauer study for $[Sb(SPh_3)]$, **24** agrees excellently with the converted value of 12.7 mm $s^{-1}$ from NQR data [22].

The crystal structures of 16 compounds among the 52 compounds whose Mössbauer data are presented here have been determined: $Sb_2O_3$, **48** (rhombic), [23], $K_2[Sb_2(tart)_2]·3H_2O$, **47**, [24]; $KSbS_2$, **51**, [25]; $K_3[Sb(ox)_3]·4H_2O$, **44**, [26]; $[Sb(S_2CNEt_2)_3]$, **30,** [20]; $[Sb(Hedta)]·2H_2O$, **37**, [27]; $[Sb(S_2COR)_3]$ (R = Et, **9**, [28]; Pr$^i$, **10**, [29]); $[Sb(S_2COEt)_2Br]$, **13**, [30]; $[Sb\{S_2P(OR)_2\}_3]$ (R = Me, **15** and Pr$^i$, **17**, [31]; Et, **16**, [32]); $[Sb(S_2PPh_2)_3]$, **21**, [33]; $Sb_2S_3$, **49**, [34]; $AgSbS_2$, **50**, [35] and $FeSb_2S_4$, **52**, [36].

$[Sb(SPh)_3]$, **24** is considered to have Ψ-tetrahedral $SbS_3E$ structure, since it's isomer shift and quadrupole coupling constant are very large as shown below.

As can be seen in Figure 1 the isomer shifts ($\delta$) and quadrupole coupling constants ($e^2qQ$) decrease dramatically as the coordination type changes from Ψ-

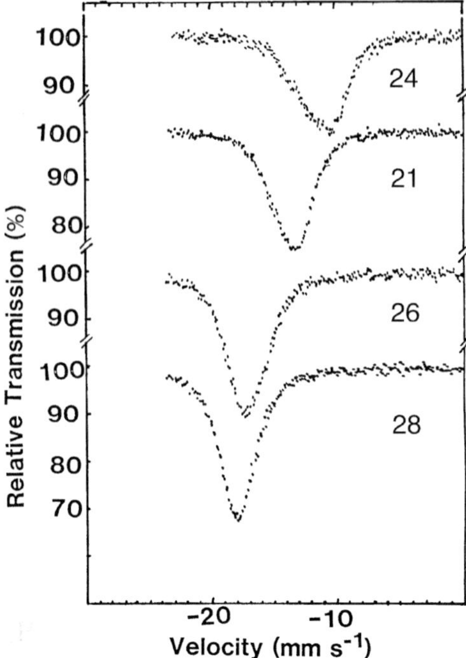


*Figure 2.* $^{121}$Sb Mössbauer spectra of antimony(III) complexes at 20 K.

tetrahedral $Sb_2O_3$, **48** to Ψ-pentagonal bipyramidal [Sb(Hedta)]·$2H_2O$, **37** (E locates in equatorial plane. Each spectrum consists of twelve superimposed absorption lines, in the Mössbauer transitions between the nuclear ground state of $I_g = 5/2$ and exited state of $I_e = 7/2$. The observed absorption peak shape with the steep decrease in the γ-ray counts in the less negative velocity region and gradual decrease in the more negative region clearly shows that the sign of $e^2qQ$ is positive, therefore the electric field gradient ($eq$) at the nucleus is negative since the electric quadrupole moment ($eQ_g$) is negative, namely −0.26 barn. This shows that the two lone-paired electrons occupy mainly $5p_z$ atomic orbital of the antimony atom. The degree of $p_z$ occupancy is larger as the $e^2qQ$ values are larger and $\delta$ values are less negative.

The velocity scale in the abscissa in Figure 2 is relative to $Ca^{119m}SnO_3$ source, so the point corresponding to 0 mm $s^{-1}$ should be moved to the right by 8.51 mm $s^{-1}$ compared to the point in Figure 1. Again the peak positions and the shape of the spectrum are different from compound to compound. The $[SbCl_6]^{3-}$ anions in **26** and **28** show nearly symmetrical absorption peaks or peak with very small values of $e^2qQ$. $[SbCl_6]^{3-}$ in $M_3[SbCl_6]$ (M=K, Rb, Cs) is essentially distorted from octahedron. The observation of nearly 0 mm $s^{-1}$ of $e^2qQ$ and the largest negative $\delta$ value of $[Co(NH_3)_6][SbCl_6]$, **28**, suggesting the very high *s* electron density at the Sb nucleus clearly proves that the electron configuration of Sb(III) ion is $[Kr](4d)^{10}(5s)^2$ without any donation to the $5p$ orbital. This is

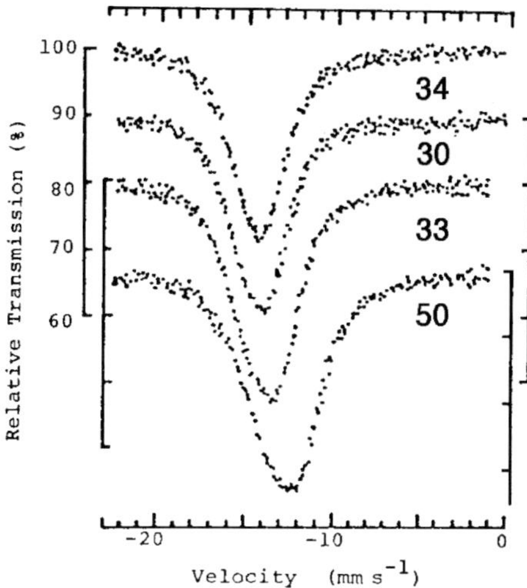


*Figure 3.* $^{121}$Sb Mössbauer spectra of antimony(III) complexes at 20 K.

just the same case as in $M_2[TeCl_6]$ (M = Rb, Cs, $NH_4$) where there is no quadrupole splitting [1].

The spectra of Figure 3 (velocity scale is relative to $Ca^{119m}SnO_3$ source) are those for the compounds of **30**, **33** and **34** with $SbS_3^{short}S_3^{long}$-S′E structure and **50** having $SbS_3^{short}S_3^{long}$E structure. The electronic effect of alkyl substituents can be seen as a shift of center of the absorption bands. The s electron density at the nucleus decrease in the order R = Bz, **34** > Et, **30** > Bu$^i$, **33** > $AgSbS_2$, **50.**

The plot of $e^2qQ$ against $\delta$ for *S*- and halogen-coordinated compounds (black symbols) is given in Figure 4 together with that for O- and N-coordinated compounds (white symbols). These data show the linear relationship between $\delta$ and $e^2qQ$ in each case.

The increased *p* character of the lone pair electrons decreases the *s* electron density at the nucleus, resulting in the rise of $\delta$ value since the nuclear radius parameter $\Delta R/R$ is negative for $^{121}$Sb. The enlarged *p* character of the lone pair also increases the value of $e^2qQ$. Thus we get linear relationship as observed here. The lone pair of [Sb(Spent$^i$)$_3$], **23** has remarkably large *p*-orbital character, in contrast to the nearly complete *s* character in $[SbCl_6]^{3-}$. Stevens and Bowen estimated the $\delta$ value of −10.6 (3) mm s$^{-1}$ for a naked $Sb^{3+}$ ion from the $\delta$ − $e^2qQ$ plot [8], which agrees well with the value of −10.78 mm s$^{-1}$ for $[Co(NH_3)_6][SbCl_6]$, **28**.

We noted the smaller quadrupole coupling constants, thus smaller 5*p* electron contribution to the lone pair in sulfur and halogen coordinated Sb(III) complexes than in oxygen and nitrogen coordinated ones. This is supposed to be due to the difference in thc bonding orbital of 2*s* and 2*p* for oxygen and nitrogen and 3*s* and 3*p*, and 4*s* (5*s*) and 4*p* (5*p*) for sulfur and chlorine, and bromine (iodine), respectively with 5*s* and 5*p* orbitals of antimony. This will be explained by the

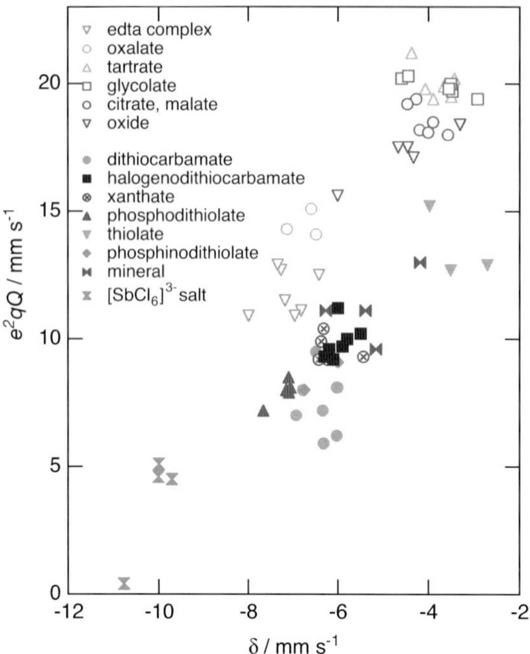


*Figure 4.* Plot of $e^2qQ$ against $\delta$ for Sb(III) compounds. Black and white symbols exhibit the parameters for S/halogen coordinated compounds and O/N coordinated compounds, respectively.

expansion of the radial part of the wave function of 5*s* and 5*p* orbital of Sb to S/Cl compared to N/O atoms. As the expansion of Sb bonding orbital proceeds, the lone pair orbital of Sb will expand.

When the lone pair has some *p* character, thus also attaining space in the coordination sphere, it causes some distortion from the ideal symmetrical structure as seen in the valence-shell electron pair repulsion theory [37]. This means that the lone pair is stereochemically active. So we can say the lone pair is stereochemically active for many types of compounds studied here except for $[SbCl_6]^{3-}$ in $[Co(NH_3)_6]$ $[SbCl_6]$. The degree of stereochemical activity depends on the type of coordination configurations and the kinds of coordinating atoms.

A schematic representation of the space of the lone pair on Sb(III) atom is shown in Figure 5, where the lone pair electrons are designated as a lobe. It is clear that Mössbauer spectra of compounds having many types of coordination polyhedra have been studied here.

Firstly from this broad investigation we can summarize the results for *S*- and halogen-coordinated compounds as follows. The values of the isomer shifts and quadrupole coupling constants, whose decrease means the decrease in the stereochemical activities of the lone pair, decrease in the order: $SbS_3E$ (Ψ-tetrahedron) in $[Sb(SR)_3]$ > $SbS_4E$ (Ψ-trigonal bipyramid, *E* is in an equatorial position) in $KSbS_2$ > $SbS_3^{short}S_3^{long}E$ (Ψ-monocapped octahedron, *E* is accommodated among the three *S* atoms at longer distances and along the $C_3$ axis) in

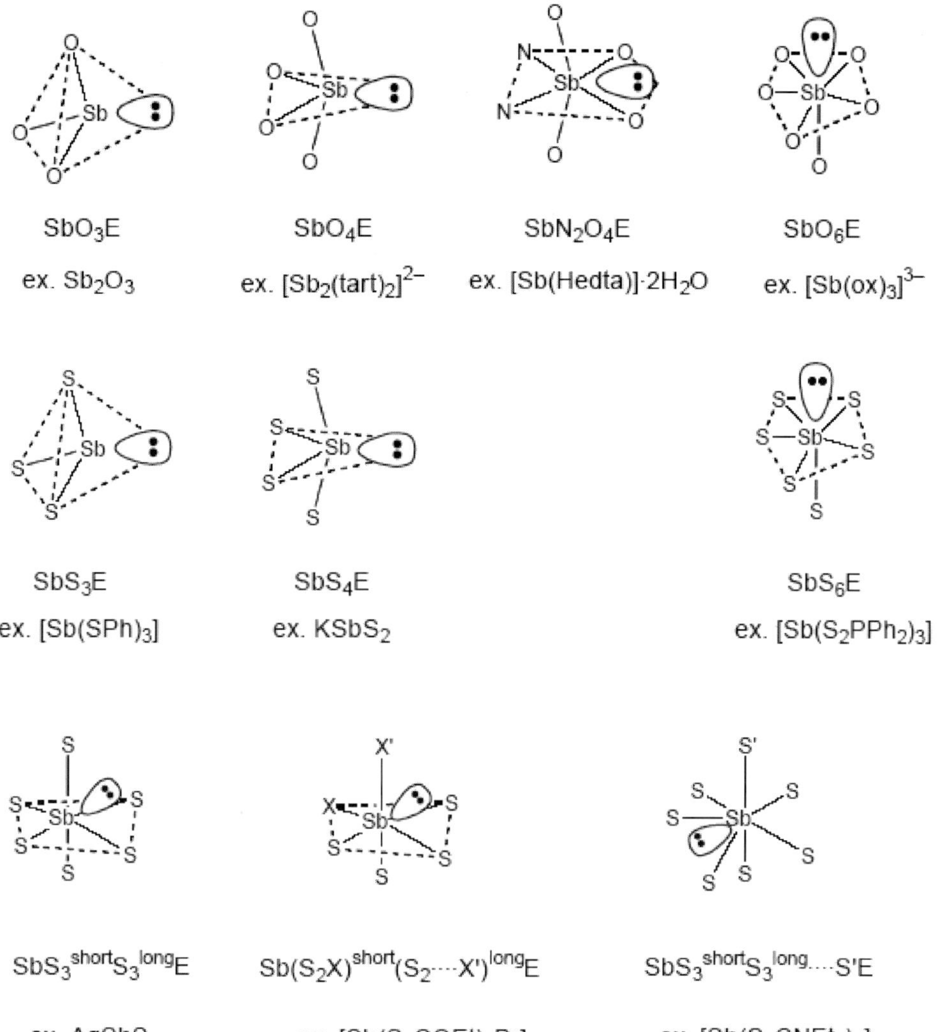


*Figure 5.* Schematic representation of coordination geometry of Sb(III) atom including a lone pair.

$AgSbS_2$ and $[Sb(S_2COR)_3]$ > $SbS_4XX'E$ (Ψ-monocapped octahedron, $Sb(S_2X)^{short}(S_2X')^{long}E$) in $[Sb(S_2COEt)_2X]$ and $SbS_5S_1E$ (Ψ-pentagonal bipyramid, $E$ occupies an apical position) in $[Sb(S_2PR_2)_3]$ > $SbS_3^{short}S_3^{long}$-$S'E$ in $[Sb(S_2CNR_2)_3]$ > $SbS_3^{short}S_3^{long}E$ (Ψ-monocapped octahedron, $E$ is accommodated among the three $S$ atoms at longer distances and along the $C_3$ axis) in $[Sb\{S_2P(OR)_2\}_3]$ > $SbCl_6$ (octahedron, $E$ is stereochemicaly non active) in $M_3[SbCl_6]$. The lone pair is accommodated only in the 5$s$ orbital in $[Co(NH_3)_6]$ $[SbCl_6]$.

Secondly in Table II Sb-S bond lengths and bond angles of the complexes having $SbS_3^{short}S_3^{long}E$ (Ψ-monocapped octahedral structure) are summarized. From $[Sb\{S_2P(OEt)_2\}_3]$, **16** down to $[Sb(S_2COPr^i)_3]$, **10** the complexes are

*Table II.* Sb–S bond lengths (Å) and bond angles (degree) in Sb(III) complexes

| | Complex | $Sb-S^{short}$ | $Sb-S^{long}$ | ratio[a] | difference | $S^{long}$-Sb-$Sb^{short}$ angle | | | Average ($x$) | $\|90 - x\|$ |
|---|---|---|---|---|---|---|---|---|---|---|
| 16 | $[Sb\{S_2P(OEt)_2\}_3]$ | 2.601 | 2.8626 | 1.087 | 0.225 | 98.8 | 97.7 | 106.2 | 100.9 | 10.9 |
| 15 | $[Sb\{S_2P(OMe)_2\}_3]$ | 2.529 | 3.005 | 1.188 | 0.476 | 103.7 | 106.4 | 108.4 | 106.2 | 16.2 |
| 17 | $[Sb\{S_2P(OPr^i)_2\}_3]$ | 2.524 | 3.015 | 1.195 | 0.491 | 111.1 | 105.9 | 102.8 | 106.6 | 16.6 |
| 9 | $[Sb(S_2COEt)_3]$ | 2.511 | 3.002 | 1.196 | 0.491 | 113.54 | | | 113.54 | 23.54 |
| 10 | $[Sb(S_2COPr^i)_3]$ | 2.508 | 3.006 | 1.199 | 0.498 | 114.79 | | | 114.79 | 24.79 |
| 30 | $[Sb(S_2CNEt_2)_3]$ | 2.581 | 2.915 | 1.129 | 0.334 | 77.3 | 156.3 | 135.1 | 122.9 | 32.9 |
| 21 | $[Sb(S_2PPh_2)_3]$ | 2.548 | 3.029 | 1.189 | 0.481 | 71.8 | 140.1 | 70.9 | 94.3 | 4.3 |
| 13 | $[Sb(S_2COEt)_2Br]$ | 2.595 | 2.982 | 1.176 | 0.387 | 79.6 | 84.2 | 112 | 91.9 | 1.93 |
| 50 | $AgSbS_2$ | 2.53 | 3.26 | 1.29 | 0.73 | 70.5 | 105.5 | 109.6 | 5.2 | 5.2 |

[a] ratio = $d(Sb–S^{long})/d(Sb–S^{short})$.

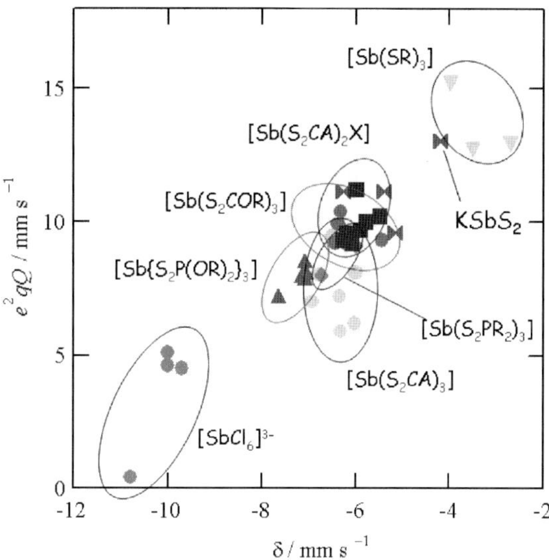


*Figure 6.* Plot of $e^2qQ$ against $\delta$ for sulfur and halogen coordinated Sb(III) complexes.

arranged in the order of increase of the distortion from octahedron. The $\delta$ value increases from $-7.15$ mm $s^{-1}$ for **16** to $-5.44$ mm $s^{-1}$ for **10**, and also $e^2qQ$ increase in the same order from 8.0 to 9.3 mm $s^{-1}$. This good relationship between $\delta - e^2qQ$ and the degree of distortion shows clearly that the $p$ character of the lone pair increases as the distortion increases.

Thirdly six $[Sb(S_2COR)_3]$ and $[Sb(S_2COEt)_2X]$ complexes, **8–14** (except **10**) having $SbS_3^{short}S_3^{long}E$ and $Sb(S_2X)^{short}(S_2X')^{long}E$ configuration, respectively have the narrow range of $\delta$ values of $-6.4$ to $-6.1$ mm $s^{-1}$ and $e^2qQ$ values of 9.2–10.4 mm $s^{-1}$. Since these values are higher than those of $[Sb\{S_2P(OR)_2\}_3]$, **15–19**, the $p$ character of the lone pair of the former compounds is larger than the latter compounds as can be seen in Figure 6. The $\eta$ values increase as the substituent becomes more bulky. The $\delta$ and $e^2qQ$ values of **10** are very similar to those of $AgSbS_2$, **50** which has the same $SbS_3^{short}S_3^{long}E$ structure. This may suggest that the structure of **10** will be similar to that of **50**.

Fourthly in the complexes $[Sb\{S_2P(OR)_2\}_3]$, **15–19** having $SbS_3^{short}S_3^{long}E$ configuration, $\delta$ values fall in the range $-(7.1–7.7)$ mm $s^{-1}$, $e^2qQ$ in the range of 7.2–8.5 mm $s^{-1}$, which show that the $s$ character of the lone pair is large. The values of $\eta$ are from 0.1 to 0.8 and increase in the order $R{=}Me<Et<Pr^i<Bu^s<Bu^i$, which correspond to the increase of the asymmetry from three fold symmetry. This in turn coincides with the order of bulkiness of the alkyl substituents.

Finally $[Sb(S_2CNEt_2)_3]$, **30** has basically $SbS_3^{short}S_3^{long}E$ configuration (Ψ-monocapped octahedron structure), but actually there is a fourth long Sb-S bond (sulfur atom of adjacent molecule is bridging) in the side of three longer Sb-S bonds, so the structure is expressed by $SbS_3^{short}S_3^{long}$-$S'E$. The $\delta$ and $e^2qQ$ values of $[Sb(S_2CNR_2)_3]$ are $-6$ to $-7$ mm $s^{-1}$ and 6–9.5 mm $s^{-1}$, respectively. $[Sb(S_2PPh_2)_3]$, **21** has a $SbS_5^{short}S_1^{long}E$ structure (Ψ-pentagonal bipyramid, $E$

occupies an apical position). Thus the structure of **21** is very similar to that of oxalato complex of **44** with $SbO_5O_1E$ configuration. Though their $\delta$ values are close, $e^2qQ$ of **21** is much smaller than that of **44** because of sulfur coordination as discussed above. $[Sb(SPh)_3]$, **24** has very large $\delta$ and $e^2qQ$ values, suggesting a very large $p$ character of the lone pair. This leads to a proposed structure Ψ-tetrahedral $SbS_3E$. The observed $\eta$ value of 0.09 is essentially zero, so $SbS_3$ is considered to be very symmetrical in **24**.

As a conclusion the stereochemical property of the lone pair has been found to depend very much on the kinds of atoms surrounding the antimony atom and the configurations of the coordinating atoms.

## Acknowledgement

This research was supported in part by a Grant-in-Aid (No. 60470049) from the Ministry of Education, Culture, Sports, Science and Technology, Japan.

## References

1. Greenwood N. N. and Gibb T. C., *Mössbauer Spectroscopy*, Chapman and Hall, London, 1971, pp. 381–390 and 441–452.
2. Donaldson J. D., Southern J. T. and Tricker M. J., *J. Chem. Soc. Dalton Trans.* (1972), 2637.
3. Takeda M., Takahashi M., Ohyama R. and Nakai I., *Hyperfine Interact.* **28** (1986), 741.
4. Takahashi M., Kitazawa T. and Takeda M., *J. Chem. Soc., Chem. Commun.* (1993), 1779.
5. Takahashi M., Matsuura A., Fujita C., Minowa K. and Takeda M., *Hyperfine Interact.* **5** (2002), 325.
6. Takeda M. and Greenwood N. N., *J. Chem. Soc. Dalton Trans.* (1976), 631.
7. Long G. G., Stevens J. G. and Bowen L. H., *Inorg. Nucl. Chem. Lett.* **5** (1969), 799.
8. Stevens J. G. and Bowen L. H., *Mössbauer Eff. Methodol.* **5** (1969), 27.
9. Stevens J. G. and Trooster J. M., *J. Chem. Soc., Dalton Trans.* (1979), 740.
10. Birchall T., Della Valle B., Martineau E. and Miline J. B., *J. Chem. Soc. Dalton Trans.* (1971), 1855.
11. Donaldson J. D., Tricker M. J. and Dale B. W., *J. Chem. Soc., Dalton Trans.* (1972), 893.
12. Manoussakis G. E. and Tsipis C. A., *Can. J. Chem.* **53** (1974), 1531.
13. Tsipis C. A. and Manoussakis G. E., *Inorg. Chim. Acta* **18** (1976), 35.
14. Gottardi G., *Z. Kristallgr.* **115** (1961), 41.
15. Gable R. W., Hoskins B. F., Steen R. J., Tiekink E. R. T. and Winter G., *Inorg. Chim. Acta* **74** (1983), 15.
16. Sowerby D. B., *Inorg. Chim. Acta* **86** (1983), 87.
17. Gress M. E. and Jacobson R. A., *Inorg. Chim. Acta* **8** (1974), 209.
18. Martineau E. and Milne J. B., *J. Chem. Soc., A* (1970), 2971.
19. Barrowcliffe T., Beattie I. R., Day P. and Livingstone K., *J. Chem. Soc., A* (1967), 1810.
20. Raston C. L. and White A. H., *J. Chem. Soc. Dalton Trans.* (1976), 791.
21. Shenoy G. K., Friedt J. M., Malleta H. and Ruby S. L., In: Gruverman I. J., Seidel C. W. and Dieterly D. K. (eds.), *Mössbauer Effect Methdology*, Vol. 9, Plenum, New York, 1975.
22. Brill T. B. and Cambell N. C., *Inorg. Chem.* **12** (1973), 1884.

23. Svensson C., *Acta crystallogr., B* **30** (1974) 458.
24. Gress M. E. and Jacobson R. A., *Inorg. Chim. Acta* **8** (1974), 209.
25. Graf H. A. and Schaffer H., *Z. Anorg. Allg. Chem.* **414** (1975), 211.
26. Poore M. C. and Russel D. R., *J. Chem. Soc., Chem. Commun.* (1971), 18.
27. Shimoi M., Orita Y., Uehiro T., Kita I., Iwamoto T., Ouchi A. and Yoshino Y., *Bull. Chem. Soc. Jpn.* **53** (1980), 3189.
28. Hoskins B. F., Tiekink E. R. T. and Winter G., *Inorg. Chim. Acta* **97** (1985), 217.
29. Hoskins B. F., Tiekink E. R. T. and Winter G., *Inorg. Chim. Acta* **99** (1985), 177.
30. Gable R. W., Hoskins B. F., Steen R. J., Tiekink E. R. T. and Winter G., *Inorg. Chim. Acta* **74** (1983), 15.
31. Sowervy D. B., Haiduc I., Barbul-Rusu A. and Salajan M., *Inorg. Chim. Acta* **68** (1983), 87.
32. Day R. O., Chauvin M. M. and McEwen W. E., *Phosphorus Sulfur* **8** (1980) 121.
33. Begley M. J. and Sowervy, D. B., *Chem. Commun.* (1980), 64.
34. Bayliss P. and Nowacki W., *Z. Kristallgr.* **135** (1972), 308.
35. Knowles C. R., *Acta Crystallogr.* **17** (1964), 847.
36. Buerger M. J. and Hahn T., *Am. Mineral.* **40** (1955) 226.
37. Gillespie R. J. and Hargittai I., *The VSEPR Model of Molecular Geometry*, Allyn and Bacon, Needham Heights, 1991.

Hyperfine Interactions (2005) 161:113–122
DOI 10.1007/s10751-005-9173-5 

# Cationic Order in Double Perovskite Oxide, $Sr_2Fe_{1-x}Sc_xReO_6$ ($x = 0.05, 0.1$)

T. HERNÁNDEZ[1,2,*], F. PLAZAOLA[2], J. M. BARANDIARÁN[2] and J. M. GRENECHE[3]
[1]*Facultad de Ciencias Químicas, UANL, Av. Pedro de Alba s/n C. P., 66400 NL México City, Mexico*
[2]*Dpto.Electricidad y Electrónica, Fac. Ciencias (Leioa), Universidad del País Vasco, Apartado 644, 48080 Bilbao, Spain*
[3]*Laboratoire de Physique de l'Etat Condensé, UMR CNRS 6087 Université du Maine, Av. Olivier Messiaen, 72085 Le Mans Cedex 9, France*

**Abstract.** We have synthesized by sol–gel method the following polycrystalline double perovskite samples: $Sr_2Fe_{1-x}Sc_xReO_6$ ($x$ = 0, 0.05, 0.1). The results of the Rietveld refinements presented single double perovskite phases with orthorhombic symmetry for the system $Sr_2Fe_{1-x}Sc_xReO_6$, the differences in atomic radii between $Fe^{3+}$ and $Sc^{3+}$ cause a lowering in symmetry with respect to the parent $Sr_2FeReO_6$ tetragonal compound. The Curie temperatures are found at about 426 and 436 (±5) K for $Sr_2Fe_{0.9}Sc_{0.1}ReO_6$ and $Sr_2Fe_{0.9}Sc_{0.05}ReO_6$, respectively. The Mössbauer spectra measured at 77 K show complex hyperfine structures resulting from different magnetic contributions at $Fe^{3+}$ sites; the average hyperfine field is estimated 50 T and the isomer shift at 0.5 mm/s. At room temperature an intermediate valence state for Fe is also observed.

**Key Words:** double perovskite Fe compounds, Ferrimagnetism, Mössbauer spectrometry.

## 1. Introduction

It is well established since the sixties that many oxides of formula $A_2BB'O_6$ crystallize with double perovskite structure. Here A (Sr, Ca, Ba) is an alkaline-earth cation divalent and B and B′ are different transition metals like Fe, Mo, W, Re sixfold coordinated to oxygen [1, 2]. Great interest in this type of compounds is due to semi-metallic behavior combined with high Curie temperatures, providing thus technological potential in the spin polarizers for example [3]. The B B′ cations in the double perovskite structure can have different degrees of disorder. The factors that control its location are charge differences, ionic radii, geometry of coordination of the cation and relationship of size of cations A/B. However, the most influential parameter is the difference in charge. Indeed, this effect establishes mainly the limit between the ideal cubic symmetry of an

* Author for correspondence.

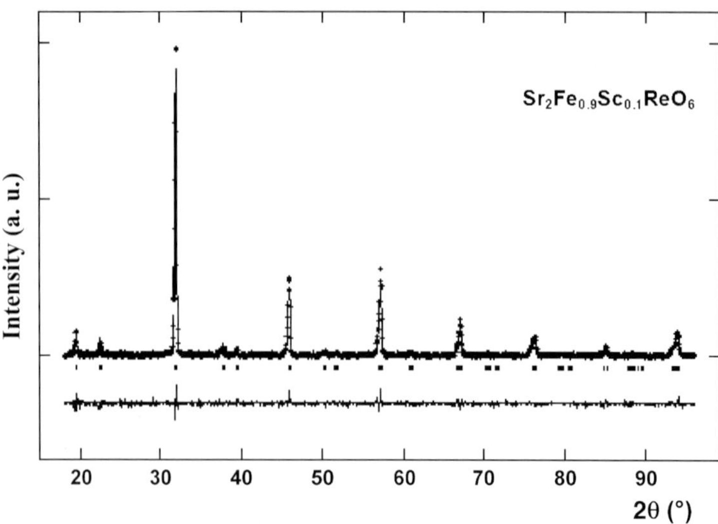


*Figure 1.* Rietveld plot from room temperature X-ray powder diffraction data for $Sr_2Fe_{0.9}Sc_{0.1}ReO_6$ *Immm*.

ordered rock-salt type superstructure and the disordered structure [4]: it has been found that the small values of ionization potentials of cations B favors its ordered structure [5].

The only magnetic interactions to be considered in the double perovskites $A_2BB'O_6$ are the B–O–B′ superexchange where the angle is exactly 180°: it is this way an ideal structure for the study of this kind of interactions between different cations [6].

As it was already mentioned, among the factors that influence the double perovskite structure, the most important one is the charge difference between cations B and B′. Indeed, disordered structures are frequently observed when the charge difference between B and B′ exceeds 2 [4]. $^{57}$Fe Mössbauer spectrometry is a very appropriate technique to distinguish different Fe sites and to determine their oxidation states, providing thus the degree of order–disorder in the case of perovskites [7–9]. In addition, localized and delocalized iron states can be evidenced and the influence of cation B′ in the environment of the nucleus of Fe as well [10].

To understand some of the phenomena of local order, structural and magnetic homogeneity, we studied the electric and magnetic nature by Mössbauer spectrometry on two samples: $Sr_2Fe_{1-x}Sc_xReO_6$ ($x = 0.05, 0.1$).

## 2. Experimental

The sol–gel method was used as precursor for obtaining the double perovskites. Indeed, this method improves the degree of homogeneity and

*Table I.* Refined cell parameters and results from magnetisation measurements for the $Sr_2Fe_{0.95}Sc_{0.05}ReO_6$ and $Sr_2Fe_{0.9}Sc_{0.1}ReO_6$ compounds

| | $Sr_2FeReO_6$ [15] | $Sr_2Fe_{0.95}Sc_{0.05}ReO_6$ | $Sr_2Fe_{0.9}Sc_{0.1}ReO_6$ |
|---|---|---|---|
| Space group | I4/*mmm* | *Immm* | *Immm* |
| a (Å) | 5.5771(2) | 5.5727(1) | 5.5759(6) |
| b (Å) | | 5.5803(1) | 5.6089(7) |
| c (Å) | 7.8930(3) | 7.8690(1) | 7.8794(8) |
| V ($Å^3$) | 243.74(2) | 244.70(5) | 246.42(8) |
| $T_c$/K | 410 | 435 | 426 |
| $M_s$ (300 K) at 0.5 T/(emu $g^{-1}$) | 21.2 | 6.9 | 5.8 |
| Ordering (%) | 92(1) | 74(1) | 78(1) |

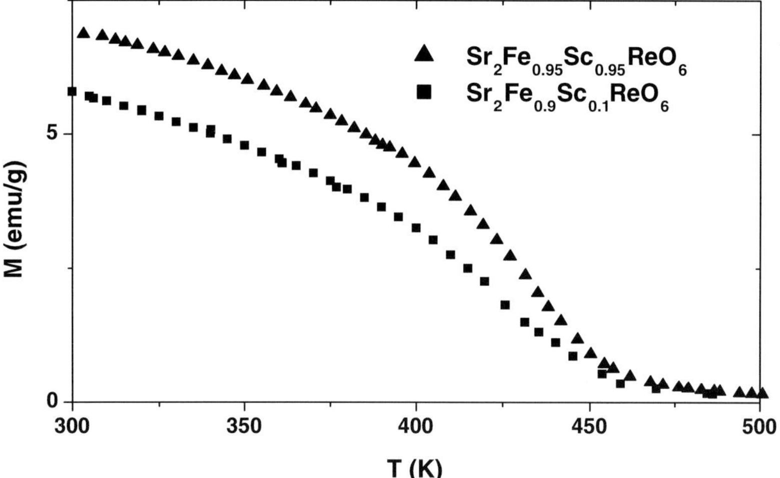


*Figure 2.* Magnetisation measurements for the studied compounds in a 0.5 T field.

cationic order, a decisive factor in the resulting properties of the double perovskites. The starting reagents were $SrCO_3$, $Fe(NO_3)_2{\cdot}9H_2O$, $Sc_2O_3$, $^{57}Fe_2O_3$ and metallic Re, all them of analytic degree. Stoichiometric amounts were dissolved in a citric acid solution to which ethylene glycol was added. Once obtained the polymer that contains the metallic ions, the solution was dried slowly in a bath of sand until the formation of a resin, the residual obtained was fired at 1300°C in Ar atmosphere. The first crystallographic characterization at room temperature of these phases was performed by X-ray powder diffraction analysis using a STOE/STADI-P powder diffractometer with Ge(III) monochromator working with $CuK\alpha_1$ radiation, in an interval of $5 < 2\theta < 120$°C with a step of 0.02°. Magnetic measurements were carried out in a Faraday balance and a SQUID magnetometer MPMS-7. Mössbauer spectrometry was carried out in transmission geometry using a conventional spectrometer of constant acceleration with a source of $^{57}$Co–Rh. Mössbauer spectra were recorded at different

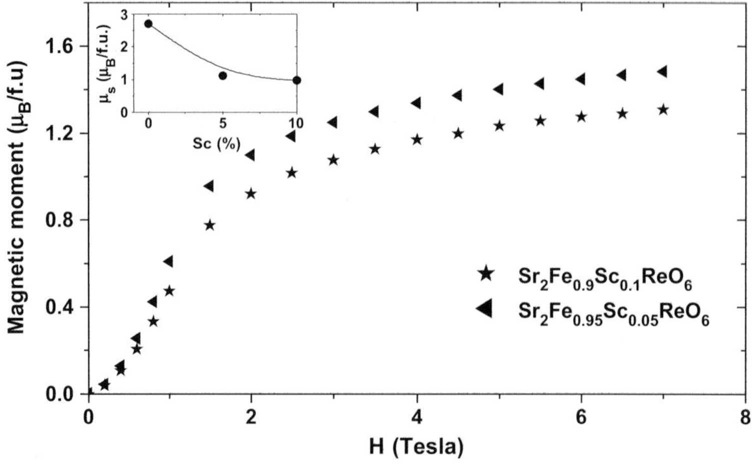


*Figure 3.* Magnetisation curves as a function at 10 K for all compounds.

temperatures using either a cryofurnace or bath cryostat. The velocity was calibrated using an Fe foil. The spectra have been adjusted using the program MOSFIT [11].

## 3. Results

As shown in Figure 1, all Bragg peaks of X-ray pattern obtained on $Sr_2Fe_{0.9}Sc_{0.1}$ $ReO_6$ correspond to a perovskite structure while the two peaks at 19° and 38° are characteristic of B B′ order superstructure. The same situation occurs for the second sample ($x = 0.05$). One can to conclude an order of $Fe^{3+}/Sc^{3+}$ and $Re^{5+}$ alternating sites in the system $Sr_2Fe_{1-x}Sc_xReO_6$ $x = 0.05$, 0.1. Indeed, the intensity of these two peaks is, therefore, related to the cationic order/disorder in the double perovskite. The structure of $Sr_2Fe_{1-x}Sc_xReO_6$ was solved using the Rietveld method [12] with orthorhombic *Immm* symmetry (the data are listed in Table I). The reliability factors are $R_{wp}$ (%): 3.5 and 4.28; $R_p$ (%): 2.7 and 3.24; and $\chi^2$: 1.16 and 1.69 for $Sr_2Fe_{0.95}Sc_{0.05}ReO_6$ and $Sr_2Fe_{0.9}Sc_{0.1}ReO_6$, respectively.

The orthorhombic *Immm* symmetry is consistent with the difference in size of the ions $Fe^{3+}$ (64.5 pm) and $Sc^{3+}$ (74.5 pm), while the volume of the unit cell increases with Fe substitution for Sc as expected when introducing a larger cation in the octahedral sites [13]. The results of the Rietveld analyses suggest 74(1)% and 78(1)% ordering in the orthorhombic phases of $Sr_2Fe_{0.95}Sc_{0.05}ReO_6$ and $Sr_2Fe_{0.9}Sc_{0.1}ReO_6$, respectively.

Magnetisation *versus* temperature curves for $Sr_2Fe_{1-x}Sc_xReO_6$ ($x = 0.05$, 0.1) are shown in Figure 2. They are characteristic of a ferrimagnetic behavior and the Curie temperatures are given in Table I. One observes an increase of the Curie temperature for the substituted samples but the small number of data does not allow to draw conclusions. In addition, no saturation is observed at 10 K even

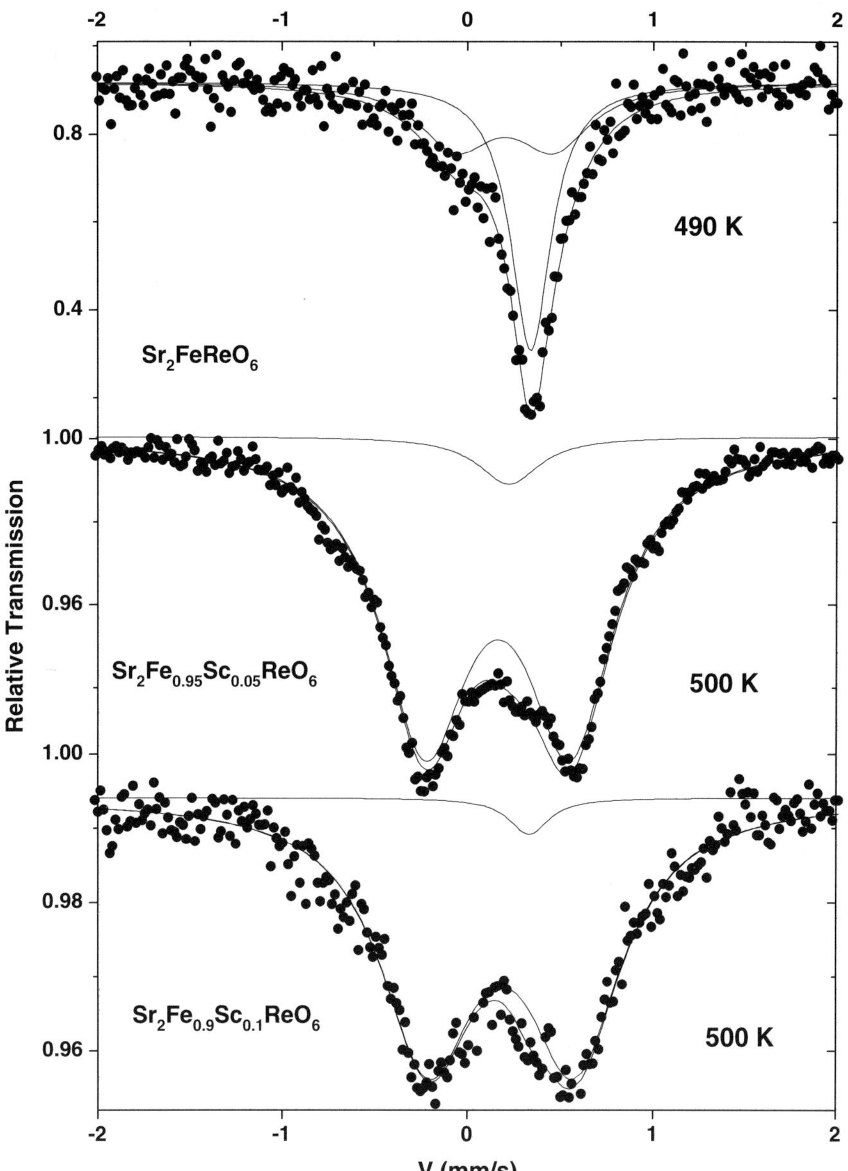


*Figure 4.* Mössbauer spectra and fitting for $Sr_2Fe_{0.95}Sc_{0.05}ReO_6$ and $Sr_2Fe_{0.9}Sc_{0.1}ReO_6$. Note that the singlet component is very small and it is inside the statistical error in the case of $x = 0.05$ and 0.1.

for 7 T, as illustrated in Figure 3 which reports the magnetisation as a function of field. The saturation magnetisation (Arrot plots [14]) diminishes drastically with the substitution of the non-magnetic $Sc^{3+}$ ion in the position of Fe, modifying thus the magnetic moment (see insert of Figure 3).

The first studies of Mössbauer spectrometry on the perovskite $Sr_2FeReO_6$ [15, 16] indicated the presence of HS $Fe^{3+}$. Gopalakrishnan *et al.* [17] proposed an

*Table II.* Mössbauer data for the $Sr_2Fe_{0.95}Sc_{0.05}ReO_6$ and $Sr_2Fe_{0.9}Sc_{0.1}ReO_6$ at 500 K

| Compound | $T$ (K) | $\delta$ (mm s$^{-1}$) | $\Delta$ (mm s$^{-1}$) | $\Gamma$ (mm s$^{-1}$) | % |
|---|---|---|---|---|---|
| $Sr_2FeReO_6$ [15] | 490 | 0.44(3) | 0.0 | 0.26(2) | 54(2) |
| | | 0.30(1) | 0.52(1) | 0.47(2) | 46(2) |
| $Sr_2Fe_{0.95}Sc_{0.05}ReO_6$ | 500 | 0.39(1) | 0.0 | 0.26(5) | 5(2) |
| | | 0.32(0) | 0.78(1) | 0.57(0) | 95(2) |
| $Sr_2Fe_{0.9}Sc_{0.1}ReO_6$ | 500 | 0.44(5) | 0.0 | 0.26(5) | 3(2) |
| | | 0.29(1) | 0.78(1) | 0.61(2) | 97(2) |

*Table III.* Mössbauer data for the compounds $Sr_2Fe_{0.95}Sc_{0.05}ReO_6$ and $Sr_2Fe_{0.9}Sc_{0.1}ReO_6$ at room temperature

| Compound | $\delta$ or $\langle\delta\rangle$ (mm s$^{-1}$) | $\Delta$/(mm s$^{-1}$) | $\Gamma$ or $\langle\Gamma\rangle$ (mm s$^{-1}$) | $\langle B\rangle$/(tesla) | % |
|---|---|---|---|---|---|
| $Sr_2FeReO_6$ [15] | 0.36(1) | 0.46(2) | 0.33(4) | | 2(2) |
| | 0.36(1) | | 0.36(4) | 43.7(5) | 98(2) |
| $Sr_2Fe_{0.95}Sc_{0.05}ReO_6$ | 0.40(1) | 0.75(1) | 0.55(4) | | 10(2) |
| | 0.41(1) | | 0.30(4) | 38.5(5) | 90(2) |
| $Sr_2Fe_{0.9}Sc_{0.1}ReO_6$ | 0.40(1) | 0.75(1) | 0.55(4) | | 13(2) |
| | 0.41(1) | | 0.30(4) | 36.6(5) | 87(2) |

intermediate state of oxidation $Fe^{2+}/Fe^{3+}$, which has been then confirmed by Blanco *et al.* [15].

The Mössbauer spectra recorded at 500 K on $Sr_2Fe_{1-x}Sc_xReO_6$ ($x = 0$, 0.5 and 0.1) are displayed in Figure 4. The asymmetrical quadrupolar spectra are fitted using two components: a single line and a quadrupolar doublet. The contribution of the single line is very small (2%–3%) for substituted samples but prevails on $Sr_2FeReO_6$ spectrum. The values of the isomer shift doublet are clearly attributed to the presence of HS $Fe^{3+}$ ions, while that of the single line (which is significantly larger) an intermediate state of valence of Fe as previously encountered in non-substituted compounds [18]. The values of isomer shift and quadrupole splitting are listed at 500 K in Table II. The rather broad linewidths of the quadrupolar doublets are due to the cationic disorder, which increases with the substitution of Sc. Similar results were found in disordered oxides where small changes in the local environment of the Fe give a distribution of quadrupolar splitting [19].

The Mössbauer spectra at room temperature of present double perovskites (Figure 5) show hyperfine magnetic structures with broad lines: the vicinity of the magnetic ordering temperature is responsible of the small quadrupolar component but the cationic disorder originates different environment of Fe sites, and consequently strongly contributes to the broadening of lines. The spectra

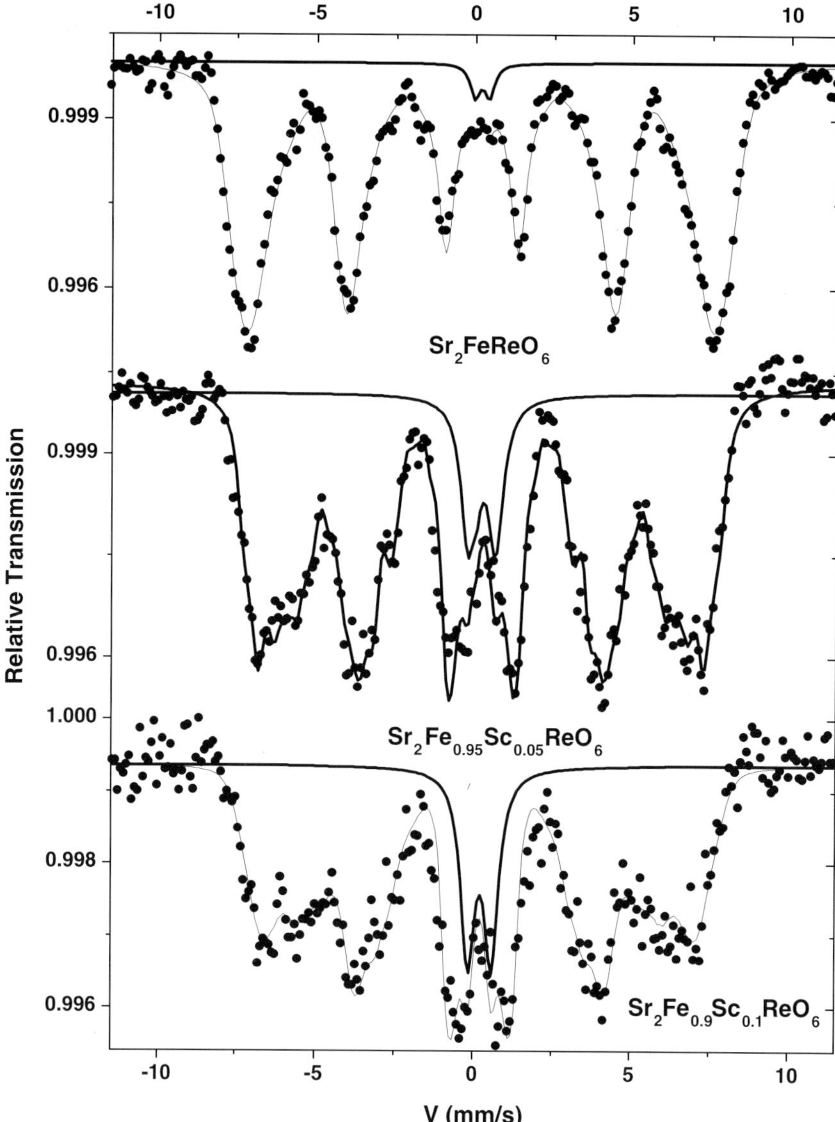


*Figure 5.* Mössbauer spectra for the compositions $x = 0.05$ and $0.1$ at room temperature.

were thus adjusted using a distribution of hyperfine fields and a small quadrupolar doublet. The values of the hyperfine parameters are given in Table III. It is important to emphasize that the two components possess the same isomer shift value to those obtained at 500 K. The Mössbauer spectra at 77 K (see Figures 6) confirm the existence of a well-developed magnetic order and present rather broad and weakly asymmetric lines due to the existence of different magnetic environments of the Fe sites within the structure. In addition, a small paramagnetic component (<5%) remains present in all the phases. Those features

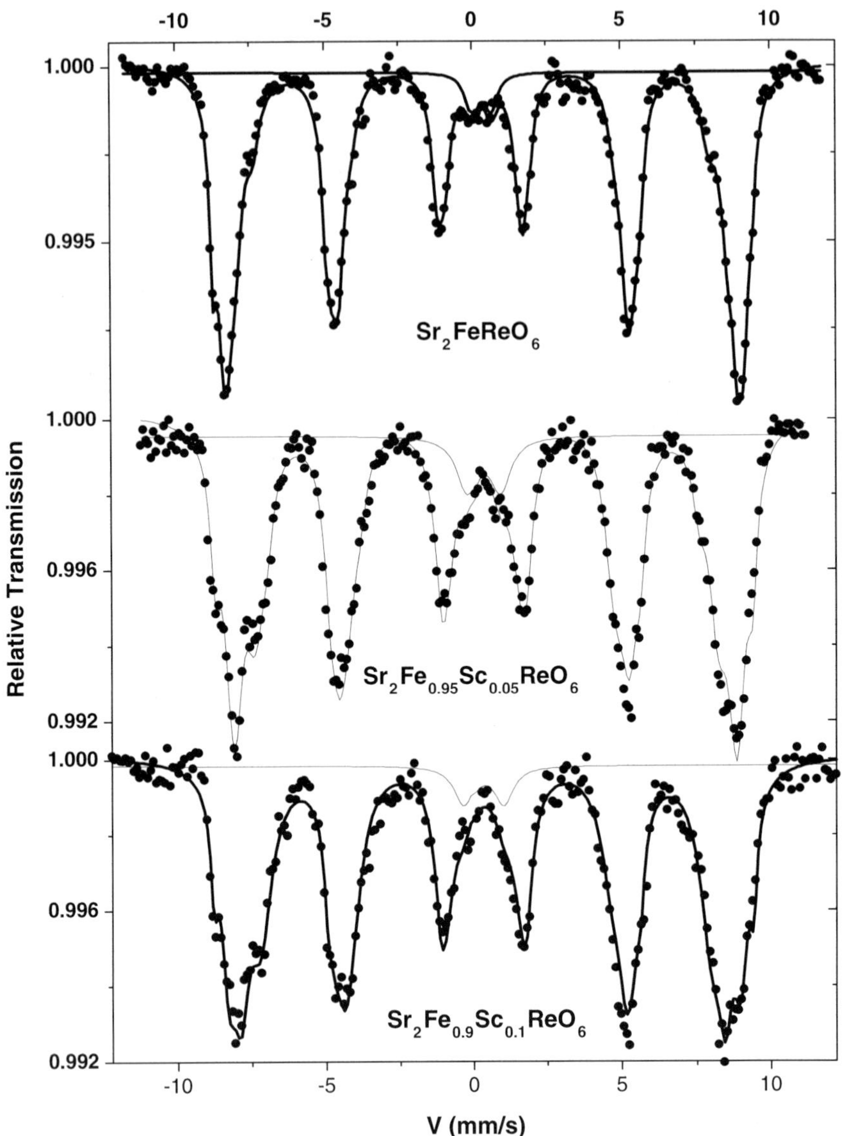


*Figure 6.* Mössbauer spectra for the samples $Sr_2Fe_{0.95}Sc_{0.05}ReO_6$ and $Sr_2Fe_{0.9}Sc_{0.1}ReO_6$ magnetically ordered at 77 K.

confirm the existence of a certain degree of disorder in the positions Fe/Re. Moreover the substitution of Sc enhances the cationic disorder. The isomer shift values remain similar and typical of HS $Fe^{3+}$ ions: one concludes thus that the intermediate state observed at high temperature disappears at low temperature, that is contrary to $Sr_2FeMoO_6$ [8, 9] for which the electron transfer is temperature independent. The values of hyperfine parameters are listed in Table IV.

*Table IV.* Mössbauer data for the system $Sr_2Fe_{1-x}Sc_xReO_6$ ($x = 0.05, 0.1$) at 77 K

| Compound | $\langle\delta\rangle$ (mm/s) | $\langle 2\varepsilon\rangle$ (mm/s) | $\langle B\rangle$ (T) |
|---|---|---|---|
| $Sr_2FeReO_6$ [15] | 0.51(1) | −0.02(1) | 52.5(5) |
| $Sr_2Fe_{0.95}Sc_{0.05}ReO_6$ | 0.53(1) | 0.05(1) | 50.9(5) |
| $Sr_2Fe_{0.9}Sc_{0.1}ReO_6$ | 0.50(1) | 0.02(1) | 50.2(5) |

## 4. Conclusion

The Mössbauer spectra provide relevant information on the electronic configuration of Fe in the present double perovskite structures. Particularly, it is clearly evidenced the existence of two Fe types: (1) HS $Fe^{3+}$ whatever the temperature is and (2) an intermediate valence state which is only observed at high temperature. The complex hyperfine structures which are observed at different temperatures are explained by the cationic Fe/Re disorder which increases with the Sc substitution. In addition, it is consistent with the results of X-ray diffraction and well explains the lack of saturation even with large magnetic fields.

## References


1. Coey J. M. D., Viret M. and von Molnár S., *Adv. Phys.* **48** (1999), 167.
2. Sleight A. W. and Ward R., *J. Am. Chem. Soc.* **83** (1961), 1088.
3. Wolf S. A., Awschalom D. D., Buhrman R. A., Daughton J. M., von Molnar S., Roukes M. L., Chtkelkanova A. Y. and Treger D. M., *Science* **294** (2001), 1488.
4. Anderson M. T., Greenwood K. B., Taylor G. A. and Poeppelmeier K. R., *Prog. Solid State Chem.* **22** (1993), 197.
5. Bokov A. A., Protsenko N. P. and Ye Z.-G., *J. Phys. Chem. Solids* **61** (2000), 1519.
6. Popov G., Lobanov M. V., Tsiper E. V., Greenblatt M., Caspi E. N., Borissov A., Kiryukhin V. and Lynn J. W., *J. Phys., Condens. Matter* **16** (2004), 135.
7. Douvalis A. P., Venkatesan M., Coey J. M. D., Grafoute M., Greneche J. M. and Suryanarayanan R., *J. Phys., Condens. Matter* **14** (2002), 12611.
8. Pinsart-Gaudart L., Surynarayanan R., Revcolevschi A., Rodriguez-Carvajal J., Greneche J. M., Smith P. A. I., Thomas R. M., Borges R. P. and Coey J. M. D., *J. Appl. Phys.* **87** (2000), 7118.
9. Greneche J. M., Venkatesan M., Suryanarayanan R. and Coey J. M. D., *Phys. Rev., B* **63** (2001), 174403.
10. Nakamura S., Ikezaki K., Nakagawa N., Shan Y. J. and Tanaka M., *Hyperfine Interact.* **141/142** (2002), 207.
11. Teillet J. and Varret F., unpublished MOSFIT Program University of Le Mans.
12. Rietveld H. M., **2** (1969), 65.
13. Shannon R. D., *Acta Crystallogr., Sect A* **A32** (1976), 751.
14. Arrott A., *Phys. Rev.* **108** (1957), 1394.

15. Blanco J. J., Hernández T., Rodríguez-Martínez L. M., Insausti M., Barandiarán J. M., Greneche J. M. and Rojo T., *J. Mater. Chem.* **11** (2001), 253.
16. Abe M., Nakagawa T. and Nomura S., *J. Phys. Soc. Jpn.* **35** (1973), 1360.
17. Gopalakrishnan J., Chattopadhyay A., Ogale S. B., Venkatesan T., Greene R. L., Millis A. J., Ramesha K., Hannoyer B. and Marest G., *Phys. Rev., B* **62** (2000), 9538.
18. Peña A., Gutierrez J., Rodriguez-Martinez L. M., Barandiarán J. M., Hernández T. and Rojo T., *J. Magn. Magn. Mater.* **254–255** (2003), 586.
19. Gibb T. C., Battle P. D., Bollen S. K. and Whitehead R. J., *J. Mater. Chem.* **2** (1992), 111.

Hyperfine Interactions (2005) 161:123–126
DOI 10.1007/s10751-005-9174-4 

# Synthesis of Calcium Telluride as a Possible Mössbauer Source

P. C. PIÑA[1,*], H. ARRIOLA[1] and F. GUZMÁN[2]
[1]*Facultad de Química, Universidad Nacional Autónoma de México, Coyoacán 04510, D. F., México; e-mail: cppqic@servidor.unam.mx*
[2]*Instituto Nacional de Investigaciones Nucleares, Toluca, Mexico*

**Abstract.** In this work the synthesis of calcium telluride has been carried out by two methods using enriched $^{128}$Te in both cases. The X-ray diffraction results of both samples correspond to the orthorhombic tricalcic telluride. Once the two samples of calcium telluride were identified, they were irradiated with thermal neutrons in a Nuclear Reactor. The results of both samples of iodine correspond to the 27.72 keV photo peak of $^{129}$I reported in 1962. The calcium telluride obtained by the two methods resulted in a possible Mössbauer $^{129}$I source.

## 1. Introduction

Two Mössbauer resonances of iodine exist, the 27.72 keV of $^{129}$I reported by Jha [1] in 1962 and the 57.5 keV of $^{127}$I observed by Barros [2] in 1964. Both resonances have been applied to studies of different iodine compounds. The $^{129}$I transition has better nuclear characteristics to produce Mössbauer Spectroscopy but it has the disadvantage that the ground state is radioactive with a half-life of $1.7 \times 10^7$ years and also difficult to obtain at commercial level.

Several matrixes have been used to produce Mössbauer sources of iodine, such as: ZnTe, SnTe, and $Mg_3TeO_6$ [3].

A possible Mössbauer source of $^{129}$I for studies in medicine, biology, chemistry, etcetera, is the tricalcic telluride, $Ca_3TeO_6$, synthesized with enriched $^{128}$Te, due to the special properties of its crystalline structure where tellurium has high coordination symmetry.

## 2. Experimental

In method [4] (Sample 1), the following reagents were used: calcium carbonate instead of calcium sulfate, enriched $^{128}$Te, dissolved in a concentrated solution of KOH and $H_2O_2$ at 100°C; later on it was placed in a porcelain capsule and

---

* Author for correspondence.

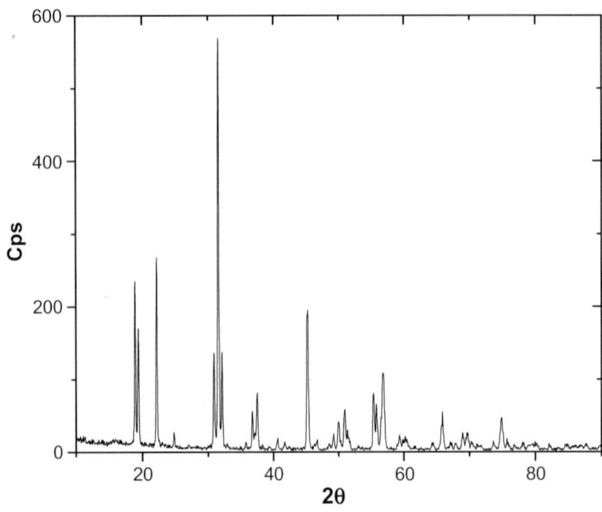


*Figure 1.* Diffractogram of telluride, method 1.

calcium carbonate was added. Finally it was placed under a thermal treatment at 900°C for 24 h.

The second method [5] (Sample 2) was carried out in a porcelain hearth with enriched $^{128}Te$ and $CaCO_3$ in the corresponding stoichiometry and nitric acid was added. Then it was heated up in a stove until total detachment of nitrous vapors and finally it was placed under a thermal treatment at 900°C for 12 h.

Once the two tellurides were obtained, they were placed in small high-density polyethylene tubes, to be irradiated in the Nuclear Reactor of the Salazar Nuclear Center, ININ, with a flux of thermal neutrons of $1.6 \times 10^{12}$ n/cm$^2$s during 2 h.

## 3. Results and discussion

The structure of the calcium orthotelluride obtained in each one of the methods was determined by X-ray powder diffraction in a Siemens Model D 5000 diffractometer using Cu K$\alpha$ radiation ($\lambda = 1.5505$ Å) and a Ni filter in an interval ($2\theta$) from 10° to 80°.

In the corresponding diffractograms for the two methods, the presence of orthorhombic tricalcic telluride is observed. There are some traces of impurities that were not identified in the product obtained by Method 1 (Figure 1). In the sample synthesized by Method 2, a diffraction peak corresponding to CaO traces is observed (Figure 2).

The difference between both structures is the intensity of the two first interplanar distances, which coincide with the orthorhombic structure, and the presence of a single peak in 2.835 Å. In the case of the calcium monoclinic telluride reported by Baglio [6], the 2.83 and 2.84 Å distances are characteristic with intensities of 100 and 95, respectively.

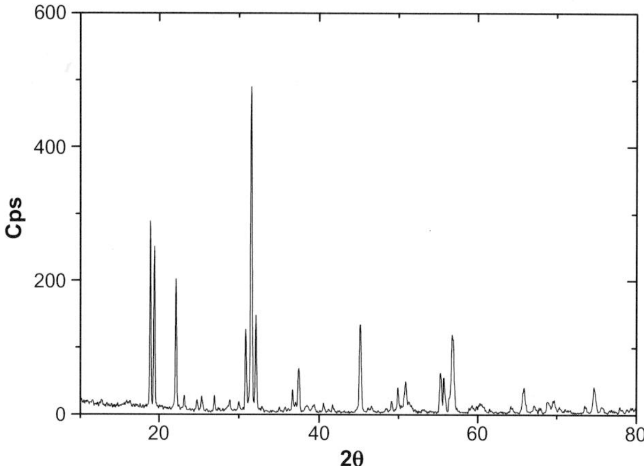


*Figure 2.* Diffractogram of telluride, method 2.

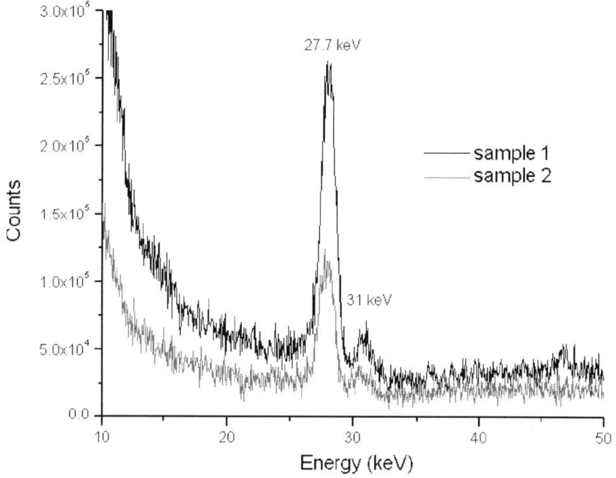


*Figure 3.* X-ray activation spectra of tellurides 1 and 2.

Once the calcium telluride was synthesized with enriched $^{128}$Te, it was irradiated with thermal neutrons from the Nuclear Reactor Triga Mark III of the National Institute of Nuclear Investigations in Salazar, to a flux of $1.6 \times 10^{12}$ neutrons/cm$^2$ s, during 2 h to produce the reaction $^{128}$Te (n, $\gamma$) $^{129}$Te.

The tellurium decays to $^{129}$I with a half-life of 70 min. The half-life of the excited state of $^{129}$I is 16.82 ns, and therefore the line width is $\Gamma = 0.59$ mm/s.

Figure 2 shows the X-ray spectrum of the resulting $^{129}$Te, where the 27.7 keV Mössbauer photopeak of the two synthesized samples is observed, although both contained small amounts of impurities. The difference in the peak intensities is only due to the difference in the irradiated quantity. Unfortunately $^{129}$I was not available, which didn't allow to run a Mössbauer spectra. However, from the X-ray 27.7 keV, it is evident that the spectrum of $^{129}$I is the appropriate one (Figure 3).

## 4. Conclusions

Orthorhombic tricalcic telluride, $Ca_3TeO_6$, was synthesized using enriched $^{128}$Te. Thermal neutron irradiation produces the Mössbauer 27.72 keV of $^{129}$I reported by Jha [1]. The calcium telluride obtained by the two methods and irradiated with neutrons resulted in a possible Mössbauer $^{129}$I source.

## References

1. Jha S., Signan R. and Lang G., *Phys. Rev.* **128** (1962), 1160.
2. Barros F. D. S. *et al.*, *Phys. Lett.* **13** (1964), 142.
3. Newnham R. E. *et al.*, Crystal Structure of $Mg_3TeO_6$, *Mat. Res. Bull.* **5** (1970), 199–202.
4. Pasternak M., A high quality source for iodine Mössbauer spectroscopy, *Research* **84** (1984), 152–154.
5. Piña P. C., *Síntesis de Ortoteluratos de Metales Alcalinotérreos*, Práctica de Química del Estado Sólido, Facultad de Química, UNAM, 2004.
6. Baglio A. Y. and Natansohn S., *J. Appl. Cryst.* **2** (1969), 252–254.

Hyperfine Interactions (2005) 161:127–137
DOI 10.1007/s10751-005-9175-3


# On the Rust Products Formed on Weathering and Carbon Steels Exposed to Chloride in Dry–Wet Cyclical Processes

K. E. GARCÍA[1,2], A. L. MORALES[1,2], C. A. BARRERO[1,2,*] and J. M. GRENECHE[3]
[1]*Grupo de Corrosión y Protección, Facultad de Ingenierías, Universidad de Antioquia, Medellín, Colombia*
[2]*Grupo de Estado Sólido, Instituto de Física, Universidad de Antioquia, A.A. 1226, Medellín, Colombia; e-mail: cbarrero@fisica.udea.edu.co*
[3]*Laboratoire de Physique de l'Etat Condensé - UMR CNRS 6087, Université du Maine, 72085, Le Mans Cedex 9, France*

**Abstract.** The rust products formed on weathering and carbon steels exposed to dry–wet cyclical processes in different chloride-rich solutions are carefully examined by means of different techniques. Special emphasis is given to the methodology of analysis of the data using 300 K and 77 K Mössbauer spectrometry and X-ray diffraction. The rust that is loosely bound to the metal surface and that it is lost during the corrosion process, for both types of steel, was found to be composed of lepidocrocite, superparamagnetic goethite, hematite, and traces of akaganeite. On the other hand, the adherent rust, which is differentiated as scraped and hit according to the way it is obtained, from both steels was found to be composed of akaganeite, spinel phase, goethite exhibiting broad distribution of particle sizes and lepidocrocite. The relative abundances of rust components for both steels were very similar, suggesting similar corrosion processes. Mass loss measurements show that the corrosion rates increases with increasing the chloride concentration. The presence of large quantities of spinel phase and akaganeite are a consequence of a corrosion process under the influence of very high chloride concentrations. Our results are useful for assessing the behavior of weathering steels where the levels of chlorides are high or in contact with sea water.

**Key Words:** atmospheric corrosion, dry–wet cyclical process, Mössbauer spectrometry.

## 1. Introduction

It is well established that the protective properties of weathering (WS) and carbon (CS) steels in a given corrosive environment depends on the characteristics of the adherent rust (AR), i.e., that rust that is bound to the metal surface. The AR can be classified as: (i) scraped, if it is obtained by scraping the steel

* Author for correspondence.

surface with a metallic brush and (ii) hit, if it is obtained by hitting the steel with a hammer. However, a better overview of the corrosion process also requires the additional characterization of the non-adherent rust (NAR), i.e., that rust that is loosely bound to the metal surface and that it is lost during the corrosion process. The analysis of this part of the rust is scarce in the literature, because in field experiments this rust is difficult to collect, but in laboratory tests this can be done easily, and thus can provide some ideas about what is happening in field experiments. García *et al.* [1] have studied the AR and NAR formed on CS coupons submitted to dry–wet cycles at different experimental conditions such as temperature, type of CS, pH, in NaCl solutions containing $1 \times 10^{-2}$ M and $1 \times 10^{-1}$ M concentrations. Spinel phase (magnetite/maghemite) was common in the AR, whereas akaganeite was present only in traces. On the other hand, the NAR was found to be composed of lepidocrocite, superparamagnetic goethite and akaganeite. The present investigation is an extension of the previous work, in order to include both WS and CS and others NaCl concentrations.

García *et al.* [2] have also studied AR and NAR formed on CS and WS exposed to total immersion tests in chloride-rich solutions. The NAR for both steels, was collected and found to be composed of lepidocrocite as the major component, and goethite and akaganeite as minor components, showing thus that these oxides are loosely bound to the rust layer . The hit and scraped AR was composed of the same iron compounds, with the additional presence of a spinel phase as a major component. In these experiments, WS was found to be about 30% more resistant to total immersion tests than CS. Interestingly is that for both steels, only 21% of the corroded iron converts completely into AR while about 45–47% of the corrosion products are NAR, and about 31–34% of the iron ions do not convert into corrosion products. No similar work on both steels exposed to dry–wet cyclical processes has been yet reported to our knowledge: this is one of the purposes of the present investigation. Dávalos *et al.* [3] have studied the effect that the rise of the temperature during the dry period had on the $SO_2$ corrosion of WS and pure iron in wet–dry cycles. The most significant result is the formation of an intermediate corrosion layer of superparamagnetic goethite only in WS.

Nasrazadani and Raman [4] proposed a process to simulate corrosion on low alloy structural steels based on a dry–wet cycle test using plain water. For continuous wetting cycles the main component is magnetite with goethite traces for CS and maghemite traces for WS. For a 2 min drying period, goethite and magnetite are the main components. When the drying period is increased up to 5 min and the number of cycles is reduced, non-stoichiometric, magnetite becomes the dominant phase, while goethite becomes the main rust component when the number of cycles increases. For the 10 min drying period the main component is maghemite, whatever the steel composition is.

Morales *et al.* [5] simulated corrosion on AISI-SAE 1008 CS, under the combined effect of sulfate, concentrations in the range of $1–2 \times 10^{-4}$ M, and chloride, concentrations in the range of 1–3 × $10^{-3}$ M, ions using dry–wet cycles. The coupons rotated continuously over a period of 41 min, 13 min of which they were immersed in the solution and the rest outside at a temperature around 45°C. For the 65 days experiment and increasing concentration of chloride/sulfate ratios from 0.8 to 1.25, the magnetite component in the scraped rust grows. However, it decreases for a ratio of 1.5 and finally it increases again for the largest ratios. Magnetic goethite, superparamagnetic goethite and lepidocrocite are also present in the AR.

A recent mechanistic modelling of wet–dry cycles on iron by Hoerlé *et al.* [6] has appeared in the literature. As a first approximation the modelling is based on the different electrochemical reactions associated with the wet and dry stages of the cycle, i.e., cathodic reactions associated with the iron oxidation during the wetting and blocking of the anodic sites during the drying.

On the other hand, the importance of doing a detailed study of corrosion products has been clearly demonstrated in the case of the protective rust formed on both WS and mild steels (MS) after 35 years of exposure to a Japanese semirural type atmosphere [7]. The authors found that, by using detailed variable temperature Mössbauer investigation, the rust on both steels are composed of goethite (major component), lepidocrocite (minor component) and traces of magnetite. Additionally, they reported that while 83% vol. of the goethite particles in the rust on mild steel are larger than about 12 nm and the rest of them are smaller than 9 nm, while the rust on WS exhibits a continuous and wide distribution down to less than 9 nm.

In this work, both the AR and the NAR formed on WS and CS exposed to dry–wet cyclical tests in different chloride-rich solutions are carefully examined by means of different techniques. Special emphasis is given to the methodology of analysis of the experimental data using variable temperature Mössbauer spectrometry and X-ray diffraction. A more extensive investigation will be reported in another document.

## 2. Experimental method

We have performed dry–wet cyclical processes on carbon steel, AISI-SAE 1020 (CS), and weathering steel, ASTM-A588 (WS). The metal composition for the A36 CS is 0.23% C, 0.40% Al, 0.49% Mn, 0.01% Ni, 0.02% Si, 0.01% S, 0.03% Cu, 0.00% Cr, and 0.01% P; whereas for A588 WS it is 0.13% C, 0.01% Al, 0.72% Mn, 0.03% Ni, 0.33% Si, 0.02% S, 0.47% Cu, 0.48% Cr, 0.02% P.

In order to check reproducibility, six coupons for each steel were employed. The CS coupons were placed into four vessels containing solutions with 1 × $10^{-1}$ M (vessel I), 6 × $10^{-1}$ M (vessel II), 5 × $10^{-3}$ M (vessel III), and 1 × $10^{-2}$ M (vessel IV) NaCl. The WS coupons were placed into two vessels with solutions of 1 × $10^{-1}$ M (vessel V), and 6 × $10^{-1}$ M (vessel VI) NaCl. The sodium chloride used was analytical grade. All coupons were previously cleaned by sandblasting, degreased with acetone and washed with water and soap and then dried in a stove. The solutions were aerated continuously during the chemical treatment to simulate the presence of oxygen. The coupons, had dimensions of 150 mm × 30 mm × 3 mm, were continuously rotated for a period of about 41 min. The samples were immersed in the solution, for 13 min; simulating the wetting period; the rest of the time, they were exposed at a temperature ranged from 40 to 80°C, the lower temperature close to the solution and the higher close to the heating source, simulating thus the drying period. These parameters were chosen according to ISO 9223, which defines time of wetness (TOW) category for different climates, in our case it is the fourth category for 31% < TOW < 60% (13 and 41 min correspond to 32% of TOW) corresponding to medium humidity climates. All solutions were changed every 48 h. We collected the rust that fell down into the solution periodically and we called it the non-adherent rust (NAR). The NAR was filtrated and washed with distilled water for removal of the excess sodium chloride, then was dried at ambient temperature, weighted, and characterized. When plain water or very mild contaminants were used, no NAR rust was collected, so the falling rust came from the metal surface and not from the solution itself. All samples were removed from each vessel after 46 days of exposure, a time long enough to produce a substantially thick adherent rust layer that we could analyzed.

Three coupons were used to calculate the average corrosion rate, expressed in μm/y per two surfaces, which was measured by means of a gravimetric test according to ISO 8407. The final solutions had pH and temperature values around 6 and 40°C respectively.

X-ray diffraction (XRD) patterns of the AR and NAR samples were obtained using a D500 Siemens diffractometer equipped with a Co (Kα) radiation; data were collected in the 10°–100° $2\theta$ range with a 0.02° step and a counting time of 10 s for step. The patterns were analyzed using a program called MAUD which combines the Rietveld method and a Fourier transform analysis, well adapted especially to describe broadened Bragg peaks [8]. This program was mainly used to derive the relative weight abundance. 300 K and 77 K Mössbauer spectra (MS) were collected in a time-mode spectrometer working in the transmission geometry using a constant acceleration drive with triangular reference signal. Calibration was achieved from a standard α-iron foil at 300 K. The spectra were analyzed using a program called MOSFIT which is based on non-linear least squares fitting procedures assuming lorentzian Mössbauer lines.

## 3. Results and discussion

At the end of the experiment, the corrosion rate for CS for two surfaces, is estimated at about 620 μm/year ($5 \times 10^{-3}$ M NaCl), 1200 μm/year ($1 \times 10^{-2}$ M NaCl), 1300 μm/year ($1 \times 10^{-1}$ M NaCl) and 3100 μm/year ($6 \times 10^{-1}$ M NaCl). The corrosion rate increases thus with increasing NaCl concentration. In the case

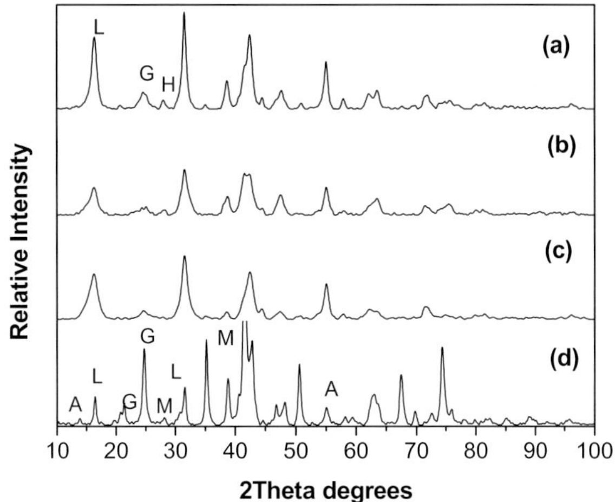


*Figure 1.* XRD patterns of the WS NAR collected after (a) 4 days, (b) 11 days and (c) 28 days of dry–wet cyclical process in vessel VI. (d) XRD pattern of the WS scraped AR collected after 46 days exposure in the same vessel.

of WS the corrosion rates are very similar or even slightly larger for the same chloride concentrations in comparison to CS, in spite of the fact that WS contains alloying elements which drive the system, in low chloride environments, to form a protective corrosion layer.

Figure 1 shows the XRD patterns for some selected samples. The upper one (Figure 1a) belongs to the non-adherent rust (NAR) collected from WS after four days of dry–wet cycles in vessel VI (0.6 M NaCl). The Bragg peaks points to the presence of lepidocrocite (L), hematite (H), goethite (G) and traces of akaganeite (A). The same iron oxides are found in the NAR collected at other exposure times (see Figure 1b and c) in both WS and CS.

Figure 2 shows the 300 K MS for NAR from WS after four days of dry–wet cycles in vessel VI. The spectrum was adequately adjusted by introducing two quadrupolar doublets, and three sextets with lorentzian line shapes, one of them with very broad lines. The quadrupolar parameters suggest the presence of a mixture of lepidocrocite, superparamagnetic (SPM) goethite and some traces of akaganeite, whereas one of the sextets is attributed to hematite, and the others to magnetic goethite. The room temperature Mössbauer spectra for the other NAR samples coming from the different vessels and at different exposure times were adjusted in a similar way. This description is in agreement with the XRD.

Figure 3 shows the 77 K MS for some NAR samples. The spectra were adjusted with one lorentzian sextet, one distribution of sextets, and one doublet, whose hyperfine parameters support previous results. The exposure time dependence of the relative abundance for the different identified iron containing phases in both steels and in different vessels is shown in Figure 4. It is possible thus to see that for WS in vessel V (0.1 M NaCl), the relative abundance of H, L and G is constant with the time, but for CS, L increases at the expense of G

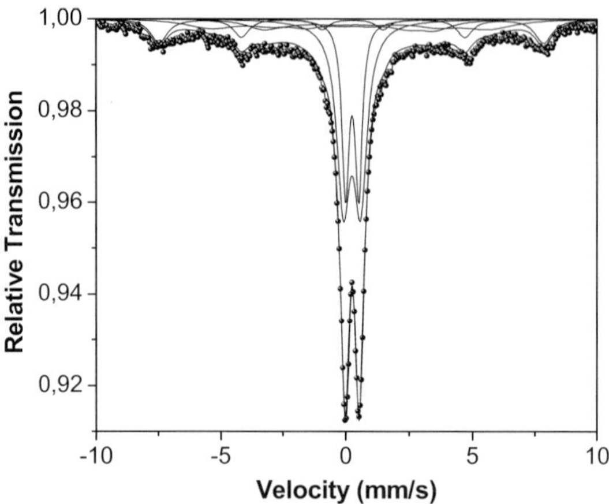


*Figure 2.* 300 K Mössbauer spectrum of the WS NAR collected after 4 days of dry–wet cyclical process in vessel VI.

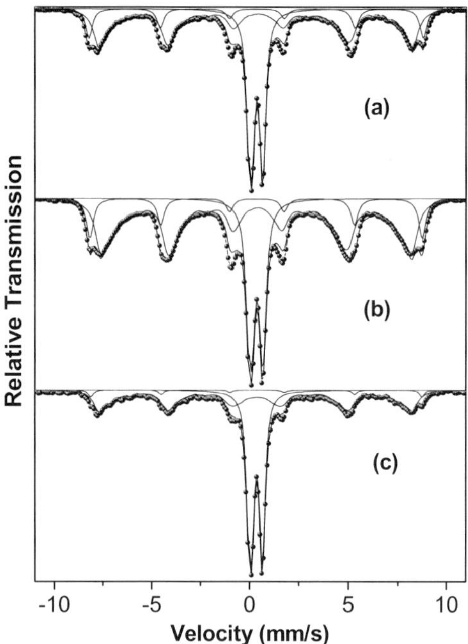


*Figure 3.* 77 K Mössbauer spectra of the WS NAR collected after (a) 4 days, (b) 11 days and (c) 28 days of dry–wet cyclical process in vessel VI.

with increasing time. At the highest 0.6 M NaCl concentration, the relative abundance of the iron phases at low exposure times is about the same for both steels, but as time goes up L increases, and H and G decreases for both steels. The observed variation in the relative abundance of the iron phases can be attributed to the different chloride concentrations and not due to the TOW,

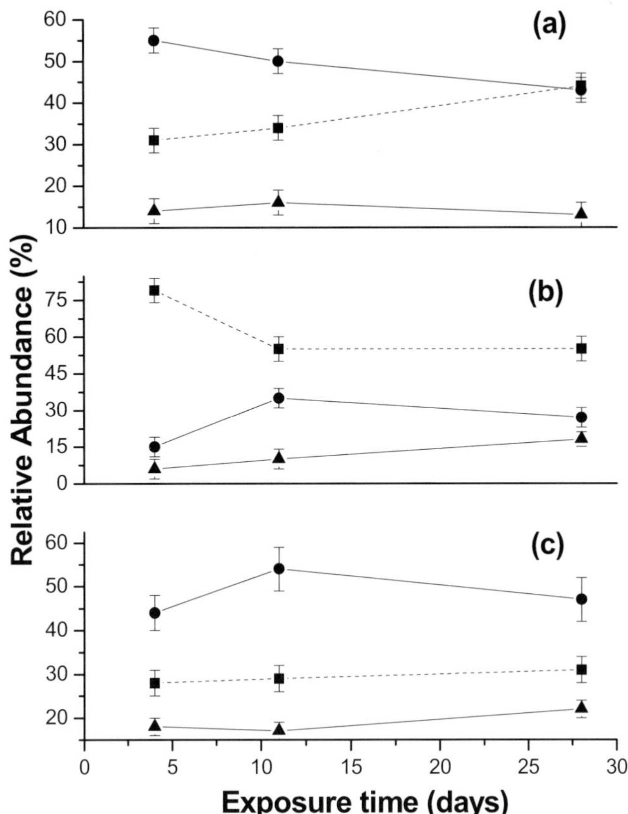


*Figure 4.* Relative abundance of hematite (*triangles*), lepidocrocite (*squares*) and goethite (*circles*) as a function of the exposure time as obtained from the relative areas in 77 K Mössbauer spectra. The results are for (a) CS in vessel I (0.1 M NaCl), (b) CS in vessel III (0.005 M NaCl), and (c) WS in vessel V (0.1 M NaCl).

because the later condition is the same for all tests. For the lowest chloride concentration, 0.005 M for CS, lepidocrocite is the dominant corrosion product at all times of the corrosion process giving a clue that different corrosion parameters do not lead to similar abundances of oxide phases.

One also observes that, for WS, G is the more abundant phase at the initial stage and L takes over at the end of the experiment. The same tendency is found for the NAR from CS at 0.1 M and 0.6 M chloride concentrations. But for smaller chloride concentrations, L remains the most abundant oxide, the content of which decreases by about 20% towards the end of the experiment. Viewing the corrosion process on the basis of the distribution of the 0.6 M NAR corrosion products, one can conclude that both steels behave in a very similar way since their distributions resemble each other. Thus, the role of the alloying elements in WS is not very effective for these chloride contents. Comparing Figure 4a and c, for 0.1 M chloride concentration less aggressive than 0.6 M, we notice that for

WS goethite is the dominant component as a result of the action of alloying elements not present in CS.

The presence of hematite in the NAR can be explained on the basis of three possible mechanisms [9]: starting (i) from poor crystalline L or (ii) poor crystalline G in acid medium, may be transformed into hematite at temperatures around 70°C. The term poor crystallinity is used here as a generic designation of two interrelated effects such as small particle size and structural disorder [10]. In our experiment the pH of our solutions were around 6–7, metal surface temperatures close to 70°C when approaching the solution, but corrosion products in the NAR exhibit poor crystallinity as is shown from XRD results; these facts may lead to H formation. (iii) A third mechanism is based on the transformation of akaganeite into hematite through a dissolution/reprecipitation process which may explain the absence of the former in the NAR. However, it is difficult to conclude from our results which mechanism prevails or maybe a combination of them might occur. These NAR products which are very active components during the corrosion process, are produced during the wet cycle: consequently, large amounts of L and G are obtained. One has to remember that the chemical formation potentials for these oxyhydroxides are very close each other [11] favoring thus to be produced simultaneously. Their characteristics are poor crystallinity, wide particle size distributions (sizes ranged from 10 nm up to 1 μm) and the presence of nanophase particles as suggested by Mössbauer spectrometry.

Figure 1d shows an XRD pattern of AR obtained by scraping the WS metal surface after 46 days of dry–wet cyclical process in vessel VI (0.6 M NaCl). The analysis of the data gives that the rust stuck to the metal surface is composed of L, G, A and a spinel phase (M). 300 K MS analysis of the B to the A site area ratio shows that this spinel phase could be either a highly oxidized magnetite or a mixture of magnetite/maghemite. No hematite was detected. The same composition was obtained for the AR obtained by hitting and scraping the metal surface of both steels, in agreement with 300 K Mössbauer results (see Figure 5).

Figure 6 shows the 77 K Mössbauer spectra of the AR from WS in vessel VI. The spectra were adjusted with one quadrupolar doublet and three distribution of sextets, assuming some constraints during the fitting procedure. The quadrupolar doublet is attributed to L while the magnetic components area assigned to A, M, G with a broad distribution of particle sizes. The relative abundances of both rust components and for both steels are very similar, suggesting thus the occurrence of similar deterioration processes and also to the formation of a non-homogeneous, not very adherent rust layer, preventing good protective properties. For these reasons, when you hit the metal sample to obtain the rust, some pieces of layer in contact with the metal surface fall off. The appearance of magnetite/maghemite and akaganeite in large amounts, for the AR, signals the occurrence of a highly deteriorating corrosion process. The spinel phase is more

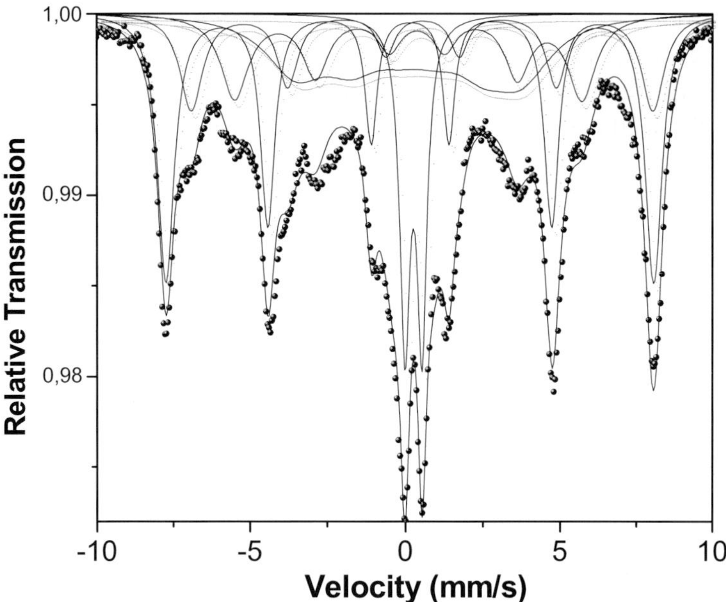


*Figure 5.* 300 K Mössbauer spectrum of the hit AR collected after 46 days of dry–wet cyclical process in vessel VI. The spectrum was adjusted with two doublets, and four sextets. The quadrupolar parameters suggest the presence of L, SPM-G and A, whereas two of the sextets are attributed to the presence of M, and the others to magnetic-G.

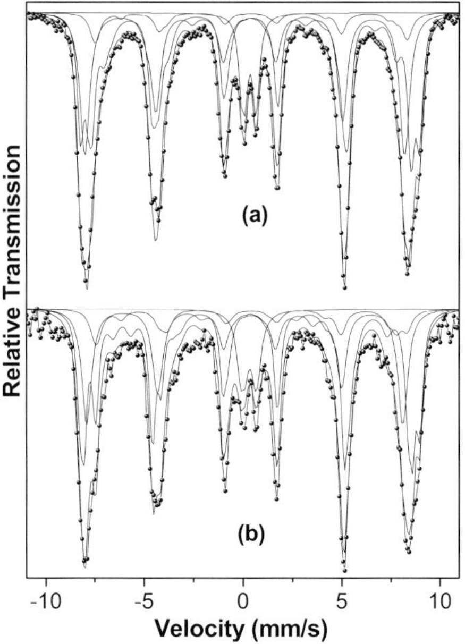


*Figure 6.* 77 K Mössbauer spectra of the AR obtained by (a) scraping and (b) hitting the WS metal surface after 46 days of dry–wet cyclical process in vessel VI.

abundant in WS while akaganeite is more abundant in CS. It is important to note that the NAR does not contain M but only traces of A: this means that these oxides remain stuck to the metal surface. It is worth mentioning that agreement between the relative abundances of the different phases as obtained from 300 K and 77 K MS and from XRD are estimated to be less than about 10%.

The results are telling us that in the experimental time window, L and G are the main oxides well formed but poor crystalline in the NAR. Also that their crystallinity is greater in the AR than in the NAR as it can be seen from Figure 1, where curves a, b, c show broad peaks attributed to L and G and curve d sharper ones. Magnetite and akaganeite are also formed during the wet cycle, the former close to the metal surface where oxygen is scarce [12] and the later is probably formed in large crystals which remain inside the rust layer and cannot flow out of it.

Bolivar *et al.* [13] found in an accelerated dry–wet test, with a 0.0001 M sulfate concentration, simulating a mild urban atmosphere and using 300 K MS, that CS developed magnetic goethite while WS did not, but in the present work both steels give magnetic goethite. In Bolivar's study, magnetite in CS was 70% of the rust while it was 23% in WS. It varies in this work from 28–43% for 0.1–0.6 M CS and 38–51% for WS. Akaganeite was not present in [13] but here a 8–20% was found. These differences in rust distribution are indicating that different corrosion mechanisms do occur. In a similar test, Morales *et al.* [5] found for CS, for a combination of 0.0001 M sulfate and 0.00174 M chloride contaminants, that the presence of magnetite was considered as a criterion of high chloride concentrations. But in our present study, magnetite and akaganeite dominate the AR rust at very high chloride concentrations.

## 4. Conclusions

The adherent and non-adherent rusts formed on weathering and carbon steels exposed to drying–wetting cycles were studied using different characterization techniques. The non-adherent rust for both steels and all chloride concentrations are composed of L, G, H, and traces of A, independent of the investigated exposure times and chloride concentrations. H is probably formed via the transformation of L, G or A in the solution. The scraped and hit adherent rusts were composed of L, M, G and A. Several differences in the crystallographic structure, relative amounts, physical and chemical properties of the iron oxides present in the different rust on both steels were evidenced. The M phase was only observed in the adherent rust, whereas H was only observed in the non-adherent rust. Akaganeite was present in the adherent rust but only traces were found in the non-adherent one. The results described here for the rust composition give keys about the possible corrosion mechanism driving the corrosion process at high concentrations of chlorides. Indications are also given about the necessity of including rust composition in mathematical models describing corrosion processes. These results are important to understand and assess the behaviour of WS and CS at high levels of chlorides.

## Acknowledgements

The financial support given by ECOS-NORD/COLCIENCIAS/ICFES /ICETEX exchange program (C03P01), COLCIENCIAS (1115-05-13661), and CODI (sustainability program for Corrosion and Protection Group 2003–2004) Universidad de Antioquia, are greatly acknowledged. We are very grateful to A.M. Mercier from Laboratoire des Fluorures of Université du Maine UMR CNRS 6010 for performing XRD measurements.

## References

1. García K. E., Morales A. L., Arroyave C. E., Barrero C. A. and Cook D. C., *Hyperfine Interact.* **148** (2003), 177.
2. García K. E., Morales A. L., Barrero C. A. and Greneche J. M., submitted to *Corros. Sci.* to appear, 2006.
3. Dávalos J., Gracia M., Marco J. F. and Gancedo J. R., *Hyperfine Interact.* **69** (1991), 871.
4. Nasrazadani S. and Raman A., *Corrosion* **49** (1993), 294.
5. Morales A. L., Cartagena D., Rendon J. L. and Valencia A., *Phys. Status Solidi, B* **230** (2000), 351.
6. Hoerlé S., Mazaudier F., Dillmann P. and Santarini G., *Corros. Sci.* **46** (2004), 1431.
7. Okada T., Ishii Y., Mizoguchi T., Tamura I., Kobayashi Y., Takagi Y., Suzuki S., Kihira H., Itou M., Usami A., Tanabe K. and Masuda K., *Jpn. J. Appl. Phys.* **39** (2000), 3382.
8. MAUD, Materials Analysis Using Diffraction by L. Lutterotti. Version: 1.84 (2002). http://www.ing.unitn.it/~luttero/maud.
9. Schwertmann U. and Cornell R. M., *Iron Oxides in the Laboratory*. 2nd Edition, VCH, Weinheim, Germany, 2000.
10. Murad E., *Phys. Chem. Miner.* **23** (1996), 248–262.
11. Arroyave C. and Morcillo M., *Trends Corros. Res.* **2** (1997), 1–16.
12. Stratmann M. and Hoffmann K., *Corros. Sci.* **29** (1989), 1329–1352.
13. Bolivar F., Arroyave C. A. and Morales A. L., *Rev. Metal.* Vol. Extr. (2003), 265–269.

Hyperfine Interactions (2005) 161:139–145
DOI 10.1007/s10751-005-9176-2


# Design and Construction of an Autonomous Control System for Mössbauer Spectrometry

A. A. VELÁSQUEZ[1,2,*], J. M. TRUJILLO[1,2], A. L. MORALES[1,2], J. E. TOBON[1,2], L. REYES[1] and J. R. GANCEDO[3]
[1]*Grupo de Instrumentación Científica y Microelectrónica, Universidad de Antioquia, A.A. 1226, Medellín, Colombia*
[2]*Grupo de Estado Sólido, Instituto de Física, Universidad de Antioquia, A.A. 1226, Medellín, Colombia; e-mail: avelasq@fisica.udea.edu.co*
[3]*Instituto de Química-Física, "Rocasolano", Consejo Superior de Investigaciones Científicas, 28006, Madrid, Spain*

**Abstract.** An autonomous control system for Mössbauer spectrometry based on two modules has been designed and built. The first module operates as a multichannel analyzer for the acquisition and storage of spectra, and the second one is a driver unit which controls and supplies the power for the velocity transducer. A microcontroller executes the digital control algorithm for the velocity transducer motion and manages the data acquisition and storage tasks. The user can monitor the system from an external PC through the serial port. A graphic interface made with the LabVIEW software allows the user to adjust digitally the control parameters for the velocity transducer motion, the channels number, to visualize as well as save spectra in a file. The microcontroller can be reprogrammed from the PC through the same serial port without intervention of a universal programmer, which allows the user to make proper software for different applications of the system. The system has been tested for linearity with several standard absorbers yielding satisfactory results. The low cost of its design, construction and maintenance make this equipment to be an attractive choice when assembling a Mössbauer spectrometer.

## 1. Introduction

The increasing use of the Mössbauer spectrometry as a magnetic and structural characterization technique of iron compounds is requiring more effort in the instrumental activity to improve performance and decrease cost by using very cheap and better integrated circuits. Since several decades ago systems based in PC and microcontrollers [1–3] carry out the multichannel analyzer (MCA) basic tasks. More recent designs of MCA make use of digital signal processing cards DSP [4] and programmable logic devices PLD [5]. The autonomous spectrometer MIMOS II [6], recently sent to Mars shows that design based in micro-controller's technology is a powerful tool, which can meet strict requirements such as: small size, high performance, high flexibility and low power

* Author for correspondence.

consumption. The former system and others as those reported by T.L Greaves and coworkers [7] implement multiple Mössbauer spectrometers, in which it is possible to take simultaneously several spectra of a sample whose properties change through the experiment, which increases the information obtained and introduces an important saving in time and cost. In a previous system built by our research group some years ago [3], the synchronization signals, the counts acquisition and the waveform signal for the velocity transducer were generated with the timer system of a PIC16C74B microcontroller, and the power stage, as well as the feedback for the velocity transducer, were performed with a commercial equipment [8]. The particular need to optimize our spectrometers by lowering costs and increasing performance motivate us to build an autonomous system with two modules: a multichannel analyzer for acquisition and storage of Mössbauer spectra and a driver unit for the control and power supply of the velocity transducer, with a microcontroller, the M68HC12B32 from Motorola family [9], as fundamental component of the system, which let us make by software some tasks that were made by hardware in the previous design, among them the generation of the control signal for the velocity transducer. The hardware and the software of the system are described in this paper.

## 2. The multichannel analyzer MCA

Its function is to perform the data acquisition, data storage as well as the generation of control signals. It includes the following elements:

**Main microcontroller M68HC12B32:** the features of this microcontroller are: 16 MHz of internal bus frequency, 16 bits arithmetic, internal timer module, serial peripheral interface, 10 bits analog to digital converter, pulse width modulator, 63 input/output lines for general purpose, 32 KB of FLASH-EEPROM memory and 1 KB of RAM memory. Its functions are: to generate the advance channel pulse through periodic interruption, to register the counts coming from the single channel analyzer, SCA, with a internal pulse counter, to store the counts in the static RAM memory, to generate the binary control signal for the velocity transducer motion and to communicate with the external PC through serial port with the RS232 communication protocol. The interrupt source is an 'output compare' flag that activates every time the free counter register of the internal timer meets the cycles number that determine the channel time of the system.

**Static RAM memory UT6264C:** this 8 KB RAM stores the counts registered in each velocity channel. It operates with a single enable line and a single Read/Write line. Four bytes of RAM are assigned for counts storage per channel, what means the maximal counts number that can be stored per channel is $2^{32} \approx 4 \times 10^9$. An external 4.8 V battery has been included in order to prevent data lost when the external power supply fails.

## 3. The driver unit

Its function is to generate the reference waveform, the control signal and the power supply for the velocity transducer motion. It includes the following elements:

**Secondary microcontroller MC6808JL3:** the MC6808JL3 [10] is a microcontroller with less hardware resources than the M68HC12B32, it operates as slave of the former one. It generates the binary code for the velocity transducer's reference waveform every time the advance channel pulse is received.

**Digital to analog converter AD669:** its function is to convert the binary waveform generated by the MC6808JL3 in the analog reference triangle signal for the velocity transducer motion. The output signal passes through an RC circuit that smoothes the steps typical of the D/A conversion.

Digital to analog converter DAC0808LCN: this element receives the binary control signal for the velocity transducer motion and converts it in an analog control signal with an 8-bit resolution.

**Power stage:** this stage contains a power amplifier STL149 of high symmetry, low noise and low offset, which supplies current to the velocity transducer control signal. Our transducer, A.S.A K-3 Linear Motor [11], consumes 800 mA DC. The power stage, built with low cost analog components, can supply up to 2 A DC.

Figure 1 shows the hardware modules of our Mössbauer spectrometer, the MCA and the driver unit have been closed in boxes.

The autonomous system has a modular structure where each sub-system executes a specific function which allows for a simpler optimization and maintenance of the system.

## 4. Software

The microcontrollers M68HC12B32 and MC6808JL3 were programmed in assembler low level language and the PC was programmed in the graphic programming language LabVIEW [12]. The MC68HC12B32 microcontroller performs primary and secondary tasks. Primary tasks require strict synchronization with the velocity transducer motion, therefore they are executed by periodic interruption. Secondary tasks do not require synchronization with the velocity transducer motion. Figure 2 shows a flow chart with the two tasks classes.

A LabVIEW graphic interface allows to the user to perform from the PC works of spectra supervision, storage the spectra in file, adjust by software of the proportional control gain, channel number and the frequency of the velocity

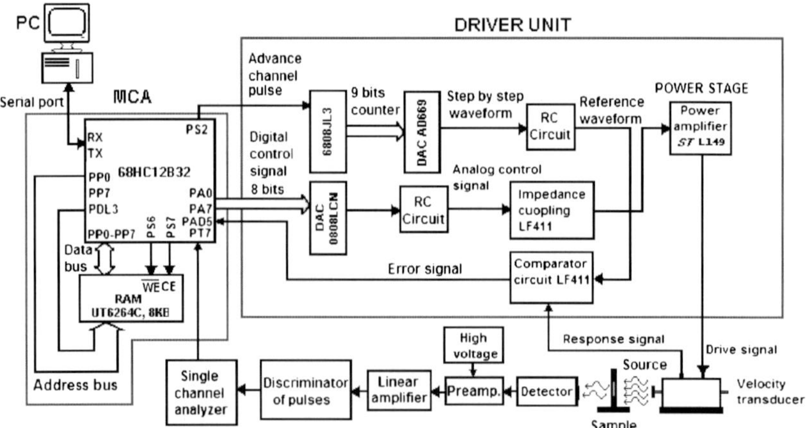


*Figure 1.* Blocks diagram of the autonomous system.

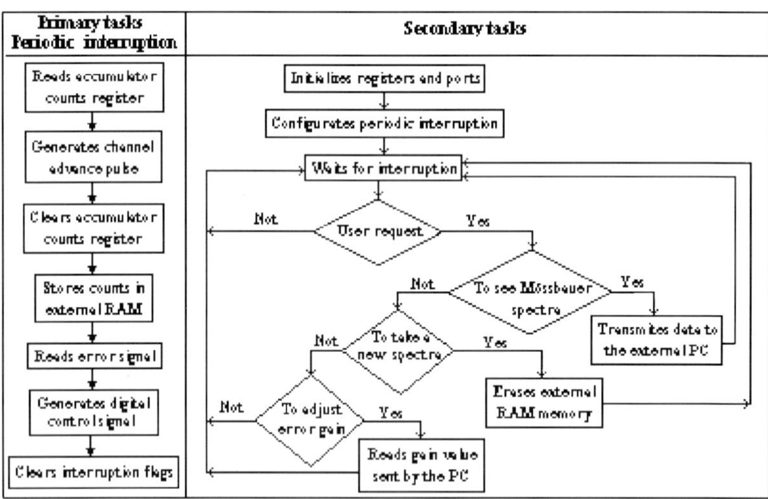


*Figure 2.* Tasks of the main microcontroller M68HC12B32.

transducer. The microcontroller M68HC12B32 can be reprogrammed from the PC through the same cable that serves for communication labors, which enables the user to program the device with other different application.

**Digital control algorithm:** the MC68HC12B32 receives the error signal, executes the analog to digital conversion with an internal ADC and then executes a proportional control algorithm that generates an 8-bit binary control signal. The former signal goes to the driver unit, where the 8-bit external DAC generates the

*Table I.* Spectral parameters obtained by fitting

| Sample | Exp. Absorption velocities: fitting (mm/s) | Standard velocities (mm/s) | Velocity deviations (%) | Correlation coefficient | Exp. width inner lines (mm/s) |
|---|---|---|---|---|---|
| $\alpha$-$Fe_2O_3$ | −8.180 ± 0.046 | −8.162 | 0.22 | 0.999998 | 0.248 ± 0.001 |
| | −4.465 ± 0.051 | −4.476 | 0.25 | | |
| | −0.954 ± 0.058 | −0.964 | 1.04 | | |
| | 1.678 ± 0.063 | 1.665 | 0.78 | | |
| | 5.177 ± 0.072 | 5.179 | 0.04 | | |
| | 8.481 ± 0.080 | 8.493 | 0.14 | | |
| $\alpha$-Fe | −5.420 ± 0.044 | −5.427 | 0.13 | 0.999998 | 0.282 ± 0.001 |
| | −3.185 ± 0.047 | −3.187 | 0.06 | | |
| | −0.954 ± 0.051 | −0.948 | 0.63 | | |
| | 0.715 ± 0.055 | 0.727 | 1.65 | | |
| | 2.963 ± 0.060 | 2.964 | 0.03 | | |
| | 5.219 ± 0.065 | 5.210 | 0.17 | | |

analog control signal and an RC circuit smoothes the steps of the conversion. The proportional control gain can be adjusted by software from the PC or by hardware with an external trimmer. Parameters adjustable by software allow that their values can be modified from a remote place when the PC operates as a host connected at network.

**Dwell time and dead time:** the MCA operates in 1024 or 512 channels, with a triangle waveform. In both cases the channel time is 162 μs. The register of counts is performed during a time equal to the channel time minus the time that takes the erased of the pulse accumulator register, which is 0.625 μs. We have then a dwell time of 161.375 μs and a dead time of 0.625 μs.

## 5. Results and discussion

Autonomous system linearity has been tested taking spectra of $\alpha$-Fe, the most accepted standard, and hematite ($\alpha$-$Fe_2O_3$), whose positions are well known. A proportional counter tube A.S.A KR-1 [8] and a $^{57}$Co/Rh radioactive source of 25 mCi were used in both cases. Table I shows the absorption positions, in velocity scale, obtained after fit the experimental data with the program CALIB [13] as well as standard positions of absorption of both samples refereed to Rhodium, relative percentage deviations in the positions, correlation coefficient of the fitting between experimental and standard positions and width of the spectra inner lines.

Figure 3 shows the Mössbauer spectra of the two samples considered.

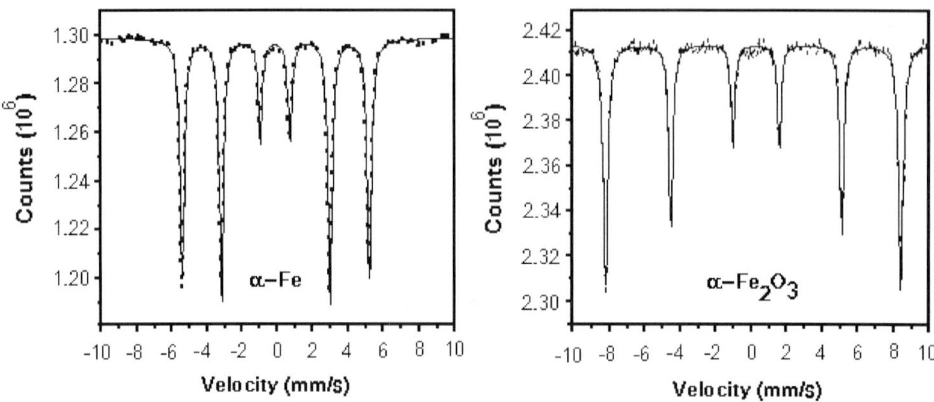


*Figure 3.* Mössbauer spectra of α-Fe and hematite taken with the autonomous system.

## 6. Conclusions

The analysis of the spectra shown and others taken in our laboratory with different iron oxides let us conclude that our system gives satisfactory Mössbauer information. Features such as low cost, modular design, fast maintenance and control parameters adjustable by software, introduce important advantages in our system.


## Acknowledgements

The authors are very grateful to COLCIENCIAS and Comité de Investigación de la Universidad de Antioquia, CODI, by the support given to this work, through the projects 1115-05-10113 and IN377CE, respectively.


## References


1. Zhou R.-J., Li F.-S. and Zhang Z.-F., *Hyperfine Interact.* **42** (1988), 1181.
2. Sánchez Asseff A. J., Bohórquez A., Pérez Alcázar G. A. and Pfannes H.-D., *Hyperfine Interact.* **110** (1997), 135.
3. Morales A. L., Zuluaga J., Cely A. and Tobón J., *Hyperfine Interact.* **134** (2001), 167.
4. CANBERRA, Multichannel Analyzers, http://www.canberra.com/pdf/Products/MCA_pdf/MCA.pdf.
5. Woodhams F. W. D. and Player M. A., *Meas. Sci. Technol.* **12** (2001), N1.
6. Klingelhöfer G., Bernhardt B., Foh J., Bonnes U., Rodionov D., De Souza P. A., Schröder C., Gellert R., Kane S., Gütlich P. and Kankeleit E., *Hyperfine Interact.* **144/145** (2002), 371.
7. Greaves T. L., Cashion J. D., Benci A. L., Jaggi N. K., Hogan C. and Bond A. M., In: *Proc. Int. Conf. on the Applications of the Mössbauer Effect*, Hyperfine Interact. (C) **5** (2001), 21.
8. Austin Science Associates, Inc, A.S.A. *Mössbauer Spectroscopy Instrumentation*, Austin, Texas, 1968.

9. Motorola Inc. 2001, M68HC12B32 Family, Advance Information, http://e-www.motorola.com/files/ microcontrollers/doc/data_sheet/M68HC12B.pdf.
10. Motorola Inc. 2000, MC68HC08JL3 Technical Data, http://e-www.motorola.com/files/microcontrollers/doc/data_sheet/MC68HC08JL3.pdf.
11. Austin Science Associates, Inc, A.S.A. Mössbauer Spectrometer: K-3, Linear Motor, Instruction Manual. Austin, Texas, 1997.
12. National Instruments. LabVIEW: Function and VI Reference Manual. USA, 1998.
13. Vandenberghe R., de Grave E. and de Bakker P. M. A., *Hyperfine Interact.* **83** (1994), 29.

Hyperfine Interactions (2005) 161:147–160
DOI 10.1007/s10751-005-9177-1


# ILEEMS: Methodology and Applications to Iron Oxides

E. DE GRAVE*, R. E. VANDENBERGHE and C. DAUWE
*Department of Subatomic and Radiation Physics, NUMAT Division, University of Gent, B-9000, Gent, Belgium; e-mail: eddy.degrave@UGent.be.*

**Abstract.** ILEEMS is the acronym for Integral Low-energy Electron Mössbauer Spectroscopy. In this variant of Mössbauer spectroscopy the low-energy electrons, $E < \sim 15$ eV, emitted by the probe nuclei in the absorber are counted as a function of source velocity. As a consequence of their low energy, the detected electrons' origin lies within a very thin surface layer with thickness of a few nanometers and consequently, ILEEMS is a useful technique to examine surfaces of Fe-containing substances. In a first part of this paper the authors briefly describe the design of a home-made ILEEMS equipment allowing the temperature of the investigated sample to be varied between 77 K and room temperature. The second part of this contribution deals with a selection of results obtained from ILEEMS spectra for various Fe oxides. In particular, the following items are covered: (1) surface versus bulk Morin transition in small-particle and near-bulk hematite, $\alpha$-$Fe_2O_3$; (2) bulk and thin-film magnetite, $Fe_3O_4$; (3) ferrihydrite, $5Fe_2O_3 \cdot 9H_20$, goethite, $\alpha$-FeOOH, and lepidocrocite, $\gamma$-FeOOH, all to some extent in relation to their morphology. Interesting and intriguing findings concerning the surface properties of these oxides were obtained and it is argued that the results encourage more systematic research in this and related fields using the surface-sensitive ILEEMS technique.

**Key Words:** iron oxides, low-energy electron Mössbauer spectroscopy, Morin transition, surface properties.

## 1. Introduction

Following resonant excitation to the 14.4 keV nuclear energy levels of $^{57}$Fe Mössbauer probes, only for 10% of the events de-excitation occurs through re-emission of gamma-rays. This feature is a consequence of the large internal conversion (~9) of the involved nuclear state, which gives rise to intense emission of electrons and secondary X-rays. The resonant electrons can be grouped into: (1) conversion electrons with an energy of 13.6 keV or higher (L and M conversion); (2) those of 7.3 keV (K conversion); (3) KLL Auger electrons with an energy of 5.4 keV; (4) LMM Auger electrons of about 580 eV; and finally (5) electrons with very low energy (<15 eV) from further decay

* Author for correspondence.

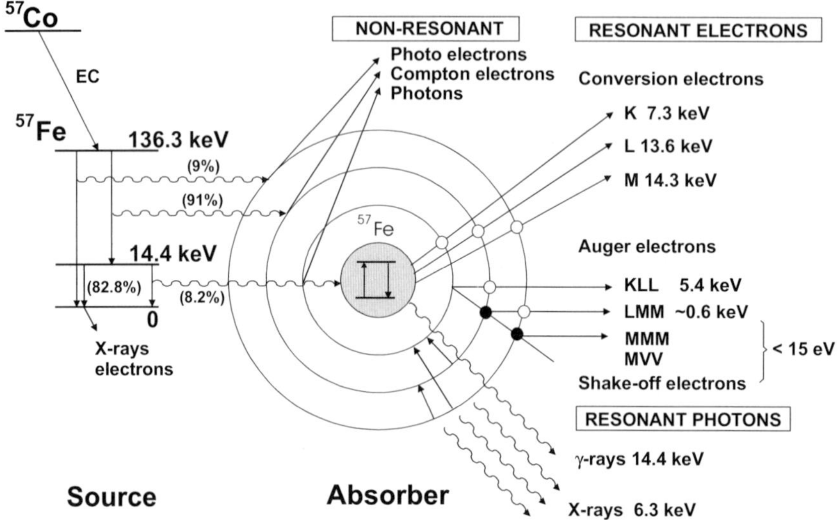


*Figure 1.* The various backscattering processes for $^{57}$Fe following resonant absorption of an incident γ-quantum.

processes (MMM, MVV, shake-off… [1]). A simplified decay diagram is presented in Figure 1. A useful variant of conventional transmission Mössbauer spectroscopy (TMS) consists of detecting the resonant electrons, which, due to their limited escape depths, yields information about the surface, which is complementary to that obtained from TMS for the bulk of the sample.

A number of electron-backscattering techniques related to the Mössbauer effect have been developed and applied over the past three decades. Most of the readers are familiar with the CEMS variant, short for Conversion Electron Mössbauer Spectroscopy. In its most simple version, using a gas-flow proportional counter [2–4], CEMS involves all electrons of relatively high energy. As such, the probing depth of the CEMS method is relatively high, i.e., 200–300 nm. More sophisticated versions of the CEMS make use of high-resolution electrostatic or magnetic electron spectrometers [5–7]. These spectrometers enable to detect electrons with energies within a certain, variable and selectable range. Because these energies are essentially related to the spatial depth of the electrons' point of origin in the surface [8, 9], they provide means to extract information about successive layers that constitute the macroscopic sample surface. In so-called DCEMS (Depth-selective Conversion Electron Mössbauer Spectroscopy) the 7.3 keV K-conversion electrons are counted energy-selectively in intervals typically of the order of 0.1 keV down to 6.5 keV. Also L-conversion electrons in the range 13.6–8.7 keV have been probed in this way [10]. Commonly, a depth resolution of a few nanometers has been obtained from the DCEM spectra collected for different energy settings on the spectrometer.

The detector used in an electron spectrometer is usually either a multi-channel plate or a channeltron. The latter is also appropriate for integral (no energy discrimination) detection of electrons [11]. Typically, its efficiency is highest for electron energies of a few hundreds of electron volt [12], which makes LMM-Auger electrons (~0.6 keV) the most efficient candidates for detection. But also for K-conversion electrons (7.3 keV), the channeltron's efficiency remains sufficiently high for it to be a useful detector. For much lower energies, however, the efficiency quickly drops to low, unworkable values, unless the electrons are given some additional energy of a few hundreds of electron volt by applying an electric field between the sample and the input of the detector. As such, the total energy of the escaped low-energy electrons is boosted to values that fall within the range for yielding optimal efficiencies of the involved detector.

Applying positive voltages to the channeltron input was indeed found to improve the quality of an electron emission spectrum, in some cases even to a spectacular extent [13]. In addition, it could be concluded from experiments on natural-Fe foil coated by an $^{57}$Fe-enriched oxide layer, that more than 50% of the integral CEM spectrum is due to electrons with energies below ~15 eV [14]. Hence, by adjusting the bias voltage on the channeltron input to the proper value to reach maximum efficiency for low-energy electron detection, one obtains a workable tool for studying near-surface Fe species. This tool is called Integral Low-energy Electron Mössbauer Spectroscopy, in short ILEEMS [15, 16]. Considering the low energies of the detected electrons, the sensitivity does not extends beyond a few nanometers below the surface, and hence properties of the Fe probes in the topmost monolayers can be examined, with no significant interference from Fe probes located at deeper layers in the surface.

## 2. The present ILEEMS spectrometer

Apart from the sample-detector arrangement, an ILEEMS spectrometer is identical to a conventional apparatus for TMS. In the present set-up, sample and channeltron are housed in an aluminium vacuum chamber in which a pressure of ~$5.10^{-5}$ mbar is maintained by a turbomolecular pump. An elementary sketch of the chamber is depicted in Figure 2. The powder sample is spread out over double-sticking carbon tape, one side being affixed to a cold finger of an insert continuous-flow cryostat and hence is at ground potential. The plane of the sample is at 45° with respect to the collimated incident γ-ray beam. The channeltron is of the Galileo spiraltron type with a 2-cm diameter entrance cone. The axis of the entrance cone is at 90° to the incident γ-ray beam. A high-voltage power supply provides, through a voltage divider (not shown in Figure 2), the anode voltage for detection of the electron signal at the end of the multiplying process in the channeltron, and the cathode voltage (bias voltage) on the input of the detector. With the aid of a potentiometer, this latter voltage is adjustable. Voltage settings of 2300 and 150 V, respectively, were generally

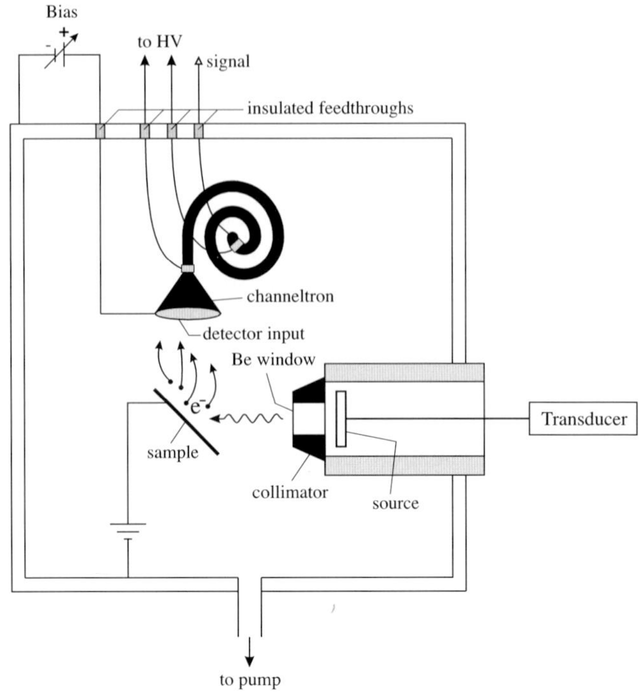


*Figure 2.* Sketch of the measuring chamber of the ILEEMS set-up.

*Figure 3.* (a) View of the inside of the ILEEMS measuring chamber from the entry port for the cryostat. On the left is the cone with Be window for the incident radiation. The channeltron is closely above it. On top of the cubic chamber on the outside is the voltage divider. (b) Picture of disassembled top plate showing the mounting of the channeltron.

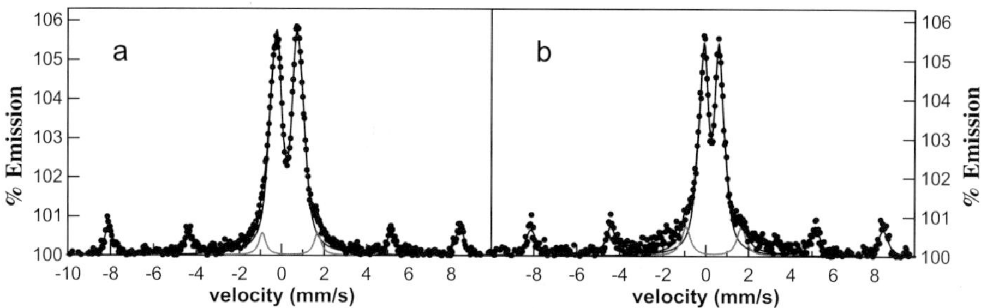


*Figure 4.* ILEEM spectrum of (a) two-line ferrihydrite and (b) six-line ferrihydrite. The presence of a magnetic component (hematite) is obvious.

found to be appropriate for oxide samples. They yield an emission of at least 500% for the outer lines of the ILEEM spectrum of a $^{57}$Fe-enriched foil, and an acquisition time for that spectrum of typically one to two hours using a reasonably active source (25 to 50 mCi). For the oxide powders, a run time of four days was on average required to obtain a statistically acceptable spectrum.

Two pictures of the partly disassembled measuring chamber and detector mounting are reproduced in Figure 3.

## 3. Applications to iron oxides

### 3.1. PHASE COMPOSITION OF Fe-OXIDE SURFACES

Figure 4 reproduces the ILEEM spectra of the synthetic six-line ferrihydrite (Fh) LC31 [17] and a natural two-line ferrihydrite (Soos, Czech Republic) obtained at room temperature (RT) using a channel width of ~0.045 mm/s and 512 channels (folded). Both species, with composition generally written as $5Fe_2O_3{\cdot}9H_20$, are poorly crystalline, with the former one showing six very broad X-ray diffraction (XRD) lines, and the latter one two, even broader lines. Mean coherence lengths, MCL, as estimated from XRD line broadenings, are of the order of 1 nm or less. TMS at RT reveals broadened, depth-asymmetric quadrupole doublets, typical of broad distributions of quadrupole splittings, arising from non-uniform Fe coordinations, which is inherent of poor crystallinity.

Clearly, the spectra in Figure 4 show the presence of an additional magnetic component. Its hyperfine parameters are almost identical for both Fh samples ($H_{hf}$ = 513 kOe, $\delta$(Fe) = 0.36 mm/s and $2\varepsilon_Q$ = −0.22 mm/s) and identify the corresponding Fe phase as hematite, $\alpha$-$Fe_2O_3$. The contribution of the sextet component to the total spectrum is (21 ± 1)% for both Fh species.

The precise structure of Fh has long been the subject of debate (see [18] for an excellent survey in that respect). Some authors have claimed that the XRD patterns can only been explained adequately if the presence of some hematite, presumably as finely dispersed grains, is postulated. Other authors have refuted this suggestion, primarily on the basis of their results from X-ray absorption

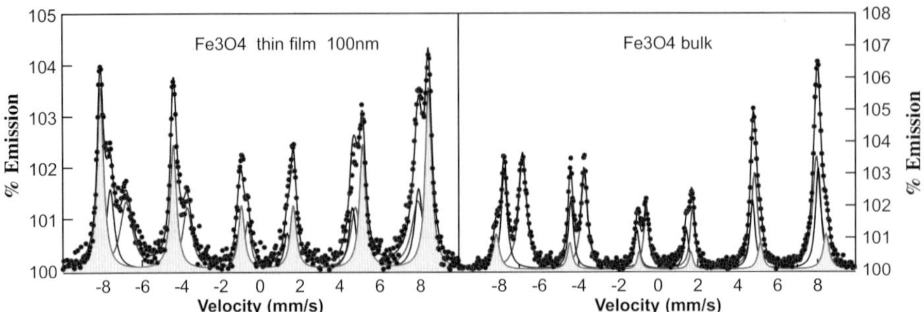


*Figure 5.* ILEEM spectra of thin film magnetite (left) and bulk magnetite. The shaded six-line pattern is due to a hematite layer.

spectroscopy (XAS). The present ILEEMS results provide prove for the hematite picture. However, the $\alpha$-$Fe_2O_3$ phase cannot be present as finely dispersed particles. If that were the case, TMS should detect these particles since they contribute by ~20% to the total iron. Hence, it must concern an extremely thin surface layer with a mass fraction against the inner Fh layers that is so small that it does not show up in TMS and XAS experiments, and hardly in the XRD patterns. The field $H_{hf}$ for the hematite layer is about 5 kOe lower as compared to the field for bulk hematite. The reduced dimension of the former may account for its reduced hyperfine field.

Subsequent ILEEMS experiments at RT on a number of different, small-particle iron oxyhydroxides, which all exhibit a doublet in TMS, have learned that also for these phases a thin surface layer of a $\alpha$-$Fe_2O_3$-like structure has formed. It concerns two lepidocrocites (Lp), $\gamma$-FeOOH, with $MCL_{020}$ values of 3.0 and 44 nm, respectively, one goethite (Gt), $\alpha$-FeOOH, ( $MCL_{111}$ = 8.5 nm) and one Al-substituted goethite with 19 mol% Al ($MCL_{111}$ = 11.5 nm). For all, the relative spectral area of the sextet component present in the ILEEM spectra is within the range 20%–30%, with no obvious correlation with the particle size, and its hyperfine field is 513 kOe. These findings suggest a rather constant thickness for the $\alpha$-$Fe_2O_3$ layer.

Not only small-particle Fe oxides have a hematite-like surface layer. Figure 5 shows the ILEEM spectra of bulk magnetite, $Fe_3O_4$, and a thin $Fe_3O_4$ film with thickness of 100 nm. The film was prepared by RF sputtering of a ceramic target on a glass substrate in an Ar plasma whose pressure was 0.5 Pa. AFM images reveal a grain size of ~20 nm. In the bulk spectrum, again a hematite component is obvious (shaded areas, $H_{hf}$ = 513 kOe, $\delta$(Fe) = 0.36 mm/s and $2\varepsilon_Q$ = −0.20 mm/s), with a contribution of 13% to the total spectrum. The relevant hyperfine parameters of the two sextets constituting the magnetite components are listed in Table I. Their values are close to those obtained from TMS and attributed to tetrahedral or A-site $Fe^{3+}$ and octahedral or B-site mixed-valence $Fe^{2.5+}$ species in the magnetite's spinel structure [19]. The ratio of the B-to A-site relative areas is ~1.6, which is lower than the ideal value of ~1.9. This finding suggests

*Table I.* Relevant Mössbauer parameters extracted from the ILEEM spectra for the A-site $Fe^{3+}$ and B-site $Fe^{2.5+}$ ions in bulk (B) and thin film (TF) magnetite

| Sample | A-site $Fe^{3+}$ | | | B-site $Fe^{2.5+}$ | | |
|---|---|---|---|---|---|---|
| | $H_{hf}$* | $\delta^{\dagger}$ | RA$^{\ddagger}$ | $H_{hf}$* | $\delta^{\dagger}$ | RA$^{\ddagger}$ |
| B | 489.2 | 0.27 | 0.33 | 458.3 | 0.67 | 0.54 |
| TF | 481.4 | 0.26 | 0.24 | 454.8 | 0.65 | 0.32 |

* hyperfine field in kOe
† isomer shift with respect to α-Fe at RT
‡ relative spectral area; the shortage from 100% is accounted for by the hematite contribution.

that, in addition to the $Fe_2O_3$ coating, the top magnetite layers may be partially oxidized [20].

The ILEEM spectrum of the thin-film magnetite displays a considerably larger relative contribution from the hematite coating, i.e., ~44%, than any of the aforementioned oxide samples. This might imply a significantly higher thickness of the hematite layer formed on this thin-film magnetite. The hyperfine field of the hematite component, 512 kOe, however, is not higher than that for the hematite layers in the Fe oxyhydroxides or in bulk magnetite, as one would expect to happen if the involved layer were featuring the same morphology, with only the layer thickness being larger. Further remarkable findings are (1) that the hyperfine field for both A- and B-site Fe species in the top layers of the magnetite thin film are significantly weaker than those of the A- and B-site Fe species in the top layers of the bulk magnetite, and (2) that for the former the ratio of the B- to A-site relative areas is further reduced to ~1.3 (see Table I), which implies a more pronounced deviation from ideal for the (average) stoichiometry of the magnetite structure in the top layers of the thin film. These are two questions that so far remain unanswered and that encourage further systematic studies along this pathway.

### 3.2. DEFORMATIONS OF SURFACE Fe SITES

Attempts were made to extract more information about the Fe coordinations in the top layers of Fh and of the two Fe oxyhydroxides, Lp and Gt. For that purpose, ILEEM spectra at RT were recorded with a reduced velocity increment per channel, i.e., ~0.013 mm/s. The as-such recorded spectrum for the two-line Fh (SOOS sample) is reproduced in Figure 6a. It was fitted with a superposition of two model-independent quadrupole-splitting distributions (QSD), one for the Fh component and one to account for the inner lines of the hematite sextet (shaded areas in Figure 6a), the results for which are of no further relevance. For the Fh QSD an adjustable, linear correlation between the quadrupole splitting $\Delta E_Q$ and the isomer shift $\delta$ was imposed. The $\Delta E_Q$ value was allowed to vary within the range 0.20–2.20 mm/s in steps of 0.025 mm/s. The as-such calculated

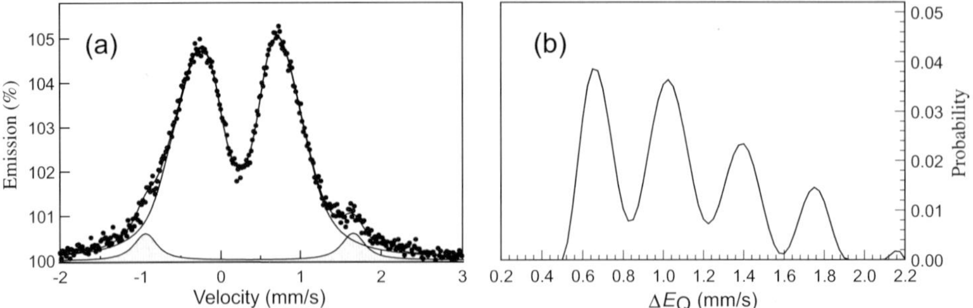


*Figure 6.* (a) ILEEM spectrum of two-line ferrihydrite recorded using a low velocity increment per channel. The shaded emission lines are the inner lines of the hematite sextet. (b) Quadrupole-splitting distribution profile fitted to the observed spectrum.

*Table II.* Quadrupole-splitting values $\Delta E_{Q,m}(i)$ and corresponding center shifts $\delta_m(i)$ versus α-Fe at RT– all in mm/s - of the $i$ relevant maxima in the QSD profiles of 6-line (LC31) and 2-line (SOOS1) ferrihydrite as derived from the ILEEM spectra at room temperature

| | | | | | |
|---|---|---|---|---|---|
| LC31 | $\Delta E_{Q,m}(i)$ | 0.546 | 0.90 | 1.44 | – |
| | $\delta_m(i)$ | 0.358 | 0.352 | 0.342 | – |
| | $p_m(i)$ | 1.0 | 0.45 | 0.14 | – |
| SOOS1 | $\Delta E_{Q,m}(i)$ | 0.660 | 1.025 | 1.40 | 1.75 |
| | $\delta_m(i)$ | 0.349 | 0.339 | 0.328 | 0.318 |
| | $p_m(i)$ | 1.0 | 0.94 | 0.60 | 0.38 |

The quantity $p_m(i)$ is the fractional probability of the $i$'th maximum with respect to the probability of the dominant maximum.

probability–distribution profile is reproduced in Figure 6b. It displays a remarkably well-resolved fine structure. Clearly, four distinct maxima are recognized in this profile. The $\Delta E_Q$ values $\Delta E_{Q,m}(i)$ of these maxima, and the corresponding $\delta_m(i)$ isomer-shift values are listed in Table II, as well as their probability value $p_m(i)$ relative to the probability $p_m(1)$ of the dominant maximum.

Similar results were obtained for the six-line Fh LC31, although the significance of the fourth maximum in the distribution profile at $\Delta E_{Q,m}(4) \approx$ 1.96 mm/s may be doubted. The numerical data for this Fh sample, obtained from the analysis performed under exactly the same conditions as those that were applied for the two-line Fh, are included in Table II. Compared to the results for the two-line Fh, the first two $\Delta E_{Q,m}$ values in the LC31 QSD are clearly and significantly shifted downwards, whilst the third maximum is at a same position and the fourth one, if real, at a higher position on the $\Delta E_Q$ scale. At this point it must be stressed that the transmission spectra of both Fh samples, recorded under conditions similar to those set for the ILEEMS, and analysed with the same input

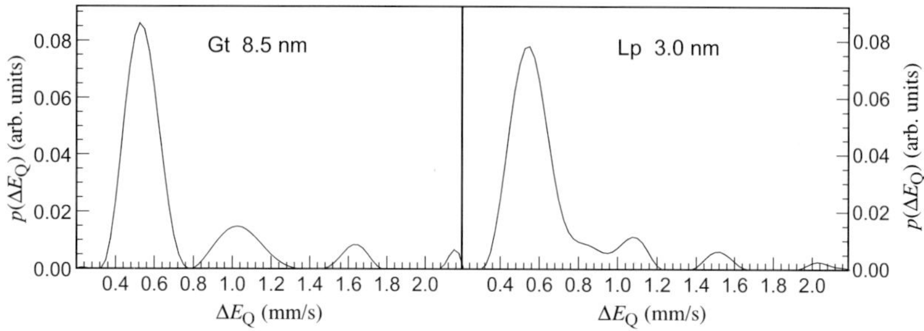


*Figure 7.* Quadrupole-splitting distribution profiles calculated from the ILEEM spectra of 8.5 nm goethite particles (left) and 3.0 nm lepidocrocite particles (right).

model for the QSD fits as that applied for the ILEEM spectra, have revealed no straightforward fine structure in the broad QSD profiles.

It is reasonable to attribute the well-resolved components in the QSD profiles to distinct Fe sites in the top monolayers of Fh. There has been dispute regarding the presence of tetrahedral Fe coordinations in these layers [18]. The present results do not support the suggestion that such sites exist. Typically, the RT centre shift $\delta$ versus α-Fe for tetrahedral $Fe^{3+}$ in oxides is ~0.26 mm/s. This value is indeed observed for the spinel oxides magnetite [19] (see also Table I) and maghemite, $\gamma$-$Fe_2O_3$ [21], as well as for many other oxide structures. No such low $\delta$ values have become apparent in the present QSD profiles. It is noticed from Table II, however, that the components with higher quadrupole splittings $\Delta E_{Q,m}(i)$, i.e., corresponding to higher distortion from octahedral symmetry, exhibit a lower $\delta_m(i)$ value. This could possibly indicate that, as site distortion increases, the octahedral coordinations gradually evolve towards more tetrahedral-like ones, some oxygens moving further away from the Fe atoms than others, but the coordination number essentially remains six.

The ILEEM spectra of the aforementioned α-and γ-FeOOH powders, measured with small velocity increments, are very similar to the Fh spectra. The derived QSD profiles for the 3.0-nm Lp sample and the 8.5-nm Gt sample are reproduced in Figure 7. Clearly, the fine structure in these profiles is significantly less pronounced than for the Fh QSD profiles, implying that the octahedral coordinations of the Fe species in the top layers of the Gt and Lp particles are less susceptible to distortions than they are for ferrihydrite.

### 3.3. SURFACE MORIN TRANSITION IN HEMATITES, α-$Fe_2O_3$

Since its discovery, the Morin transition (MT) in α–$Fe_2O_3$ has been the subject of a large variety of experimental and theoretical research papers. For a review in that respect, the reader is referred to the textbook of Morrish [22]. The transition concerns a reorientation of the $Fe^{3+}$ spins from the [111] crystallographic direction at low temperature, to the (111) basal plane at high temperature, and takes place at $T_M \approx 265$ K for pure, well-crystallized species. In the low-temperature phase, the ferric spins have a collinear antiferromagnetic ordering, the so-called AF state, whereas at high temperature, $T > T_M$, the spin structure is weakly ferromagnetic, the so-called WF state.

As a consequence of the microscopic nature of the spin–flip transition [22], $^{57}$Fe Mössbauer spectroscopy has been quite successful and effective in establishing and explaining many different aspects related to this fundamentally important magnetic phenomenon. One of these aspects concerns the relation between $T_M$ and the hematite particle size: smaller particle sizes (or, more generally speaking, poorer crystallinities) lead to decreased values for $T_M$, and to an increased temperature range $\Delta T_M = T_f - T_i$ over which the MT evolves. In this transition region, the TMS consist of a superposition of two sextet components, one arising from $Fe^{3+}$ species that belong to domains that exhibit AF ordering, and a second one arising from $Fe^{3+}$ species that belong to domains that exhibit WF ordering [23]. The two sextets are characterized by distinctive values for their hyperfine field and quadrupole shift. Commonly, as the temperature gradually decreases from above $T_M$ through the MT region, the AF contribution increases at the expense of the WF contribution. For larger particles, the WF state may eventually vanish, but for not so smaller particle sizes, for which a saturated state may persist in which both AF and WF spin states are present, with relative contributions depending on particle size. For agglomerates of particles with mean sizes of 20 nm or less, no MT is detected [24–26]. Finally, when a mixed saturation state is established, the hyperfine parameters of the AF and/or WF contributions usually differ from their values when they concern pure states, i.e. when the spectrum consists of one single sextet component. This is due to the spins being not aligned along the [111] axis for the AF state, or tilted out of the basal plane for the WF state.

In many aspects, substitution of Fe in the hematite structure by most other elements, has a similar effect on the MT as does a lowering of the particle size for non-substituted hematite. A well documented case in that respect is the Al-for-Fe substitution. With increase of Al, the MT shifts to lower temperatures and the transition range $\Delta T_M$ broadens [27]. For an Al content exceeding ~9 mol%, a MT is no longer observed and the hematite is in the WF state at all temperatures, with spins exactly in the basal plane.

It was argued in [27] that co-existence of AF and WF spin states in small-particle and/or (Al-)substituted hematites and the observed variations of their Mössbauer parameters can be qualitatively understood if one assumes that the MT in the interior of the hematite particles takes place in a higher temperature range as compared to the outer layers, implying that a given particle at a given temperature $T < T_f$ can be regarded as consisting of an interior part with a fairly uniform spin direction leaning towards the [111] crystallographic axis (AF part), and an outer part with spins tilted out of the (111) plane (WF part). On lowering $T$, the spatial expansion of the AF part grows at the expense of that of the WF

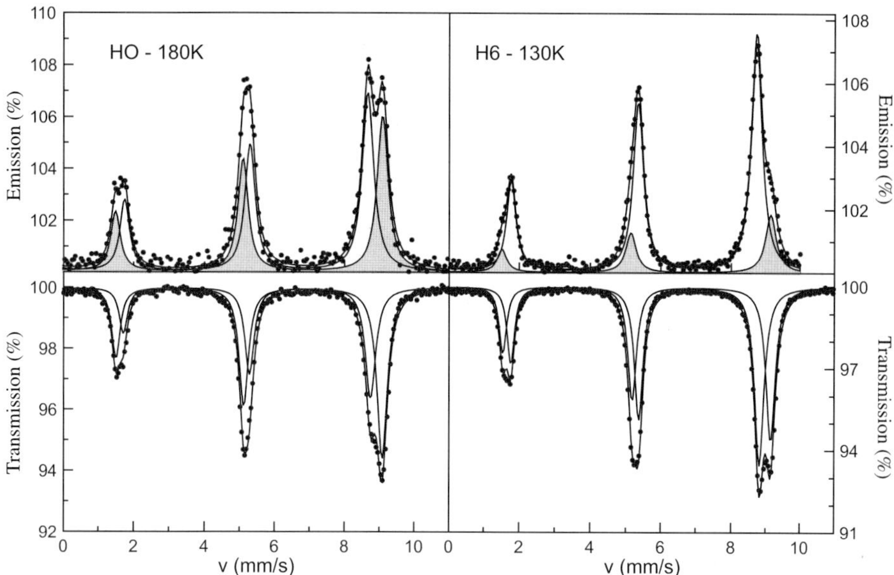


*Figure 8.* ILEEM and TM spectra of non-substituted hematite (left, H0) and hematite with a 6.5 mole% Al-for-Fe substitution (right, H6) recorded at 180 K and 130 K, respectively. For clarity, only the positive-velocity halves of the spectra are depicted. The shaded lines are due to the AF spin states.

part and simultaneously the spin orientations in both phases gradually change until the temperature has reached a lower limit $T_l$ below which a stable, possibly mixed state is established. The composition of this state, as well as the predominant spin orientations in the two coexisting sub-states depend on the average particle size and/or the Al concentration. In more recent reports, it has been suggested that the local MT would not occur as a sudden spin flip from a WF-like state to an AF-like state, but that one or more intermediate states are occupied before the final stable state is established on further lowering the temperature [28, 29].

The above pictured model of the global evolution of the MT in small-particle (Al) hematites on lowering the sample's temperature, thus implies that the local spin–flip processes for the surface Fe species differ from those for the Fe species in the core of the particles. ILEEMS seems to be a suitable tool to examine this different behaviour. Two $\alpha$–$(Fe_{1-c}Al_c)_2O_3$ samples, obtained from thermal decomposition at 500°C of a non-substituted and an Al-substituted goethite [30], and further denoted as H0 ($c$ = 0.0) and H6 ($c$ = 0.065) were considered. Their electron micrographs have shown prolate-spheroid particles with long axis of a few hundreds of nanometers and an axial ratio of 4. TMS and ILEEMS were applied as a function of $T$.

Some sample spectra are reproduced in Figure 8. For the sake of clarity, only the positive-velocity halves are displayed. Clearly, for the Al-substituted sample

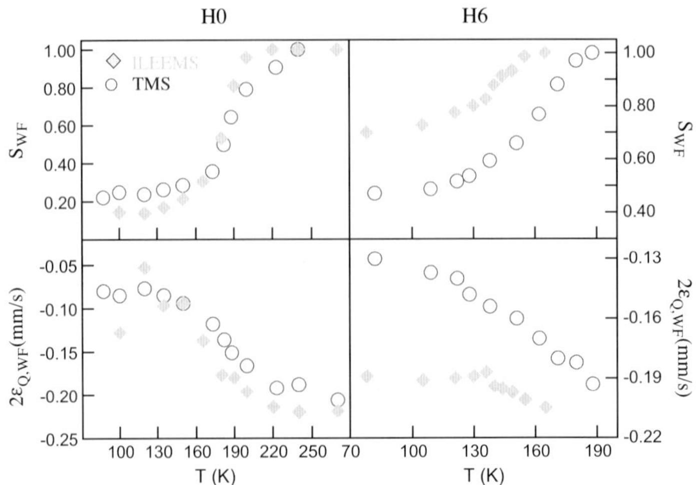


*Figure 9.* (top) Temperature variations of the relative spectral areas $S_{WF}$ of the WF contributions to the ILEEM and TM spectra of non-substituted hematite (left, H0) and hematite with a 6.5 mole% Al-for-Fe substitution (right, H6); (bottom) Temperature variations of the corresponding quadrupole shifts $2\varepsilon_{Q,WF}$.

at a given temperature in the transition region, the WF contribution in the surface (probed by ILEEMS, shaded peaks in Figure 8) significantly exceeds the WF contribution in the bulk (probed by TMS), whereas for the non-substituted sample the difference is less pronounced. This effect is further well illustrated by the temperature variations of the fractional area of the WF component, $S_{WF}(T)$, as depicted in Figure 9, top drawings. From these graphs it is seen that the temperature range over which the MT in the top layers develops, is shifted downwards as compared to the inner parts of the particles. If, as is common, the average MT temperature $T_M$ is defined as the temperature at which the WF fractional area satisfies the condition:

$$S_{WF}(T_M) = \frac{1}{2}[1 + S_{WF}(0)],$$

with $S_{WF}$ $(0)$ is the saturation value of the WF fractional area, it is estimated that the shift of $T_M$ for the surface spins is at most $\sim -5$ K for H0 and $\sim -20$ K for H6. The minute shift found for the non-substituted hematite is in agreement with earlier observations from TMS of hematite particles with $^{57}$Fe-enriched surface [31, 32].

The significant lowering of the MT in the top monolayers of small-particle Al-hematite H6 can be understood if the surface layers were richer in Al than the interior of the particles, and hence exhibit a lower $T_M$. This suggestion is supported by the ILEEM spectrum of H6 (not shown) recorded at 100 K after a heat treatment at 900°C. It consists for 97% of the WF phase, whereas the TMS of the same sample yielded only ~25% for that phase [27]. This finding may be explained by the diffusion of Al cations towards the surface regions at high temperatures. Qualitatively, an increase of the Al concentration by approximately 2% can account for the high WF fraction, an effect that was already considered possible in an earlier report [27]. Therefore, it is likely that even at 500°C (temperature of sample preparation), some enrichment of the surface has occurred.

Finally, Figure 9, bottom, shows the temperature variations of the quadrupole shifts $2\varepsilon_{Q,WF}$ of the WF components. For H6, the low-temperature saturation value for $2\varepsilon_{Q,WF}$ as derived from ILEEMS is more negative than that obtained from TMS, the former one being closer to the quadrupole shift of the pure WF spin state ($2\varepsilon_Q \approx -0.20$ mm/s). For the H0 sample, any difference has remained unclear from the present observations. However, it is likely that for this hematite too the magnitude of the (negative) ILEEMS saturation value of $2\varepsilon_{Q,WF}$ exceeds that of the TMS value, but to a lesser extent than for H6. Similar tendencies are observed for the corresponding AF quadrupole shifts (not shown), the obtained values for H0 at 100 K being 0.39 and 0.35 mm/s from ILEEMS and TMS, respectively, and 0.38 and 0.32 mm/s for H6 at 80 K. These findings can be explained in two conceivable ways: (1) the surface spin orientation with respect to the principal axis of the electric field gradient differs from the relative spin orientation in the inner parts of the grains, especially if Al is present, or (2) the magnitude of the quadrupole coupling constant in the surface layers is higher than in the deeper-lying layers. At the moment, a more complete systematic study of the ILEEMS of aluminous hematites is required to further elucidate the surface properties of the spin structure in these compounds.

## Acknowledgements

The authors wish to thank Drs. E. Murad and L. Presmanes for providing some of the samples reported on in this contribution. This work was supported by the Fund for Scientific Research, under grants No. 3G.0007.97 and No. 3G.0185.03.

## References

1. Belozerski G. N., In: *Mössbauer Studies of Surface Layers*, Elsevier, Amsterdam, 1993.
2. Terrell J. H. and Spijkerman J. J., *Appl. Phys. Lett.* **13** (1968), 1.
3. Yagnik C. M., Mazak R. A. and Collins R. L. *Nucl. Instrum. Methods* **114** (1974), 1.
4. Sawicki J. A., Sawicka B. D. and Stanek J., *Nucl. Instrum. Methods* **138** (1976), 565.
5. Parellada J., Polcari M. R., Burin K. and Rothberg G. M. *Nucl. Instrum. Methods* **179** (1981), 113.
6. Toriyama T., Asano K., Saneyoshi K. and Hisatake K. *Nucl. Instrum. Methods Phys. Res., B* **4** (1984), 170.
7. Stahl B., Klingerhöfer G., Jäger H., Keller H., Reitz T. and Kankeleit E., *Hyperfine Interact.* **58** (1990), 2547.

8. Liljequist D. and Ismail M. *Phys. Rev. B* **31** (1985), 4131.
9. Liljequist D., Ismail M., Saneyoshi K., Debusmann K., Keune W., Brand R. A. and Kiauka W., *Phys. Rev. B* **31** (1985), 4137.
10. Pancholi C. S., de Waard H., Petersen W. L. J., van der Wijk A. and van Klinken J., *Nucl. Instrum. Methods Phys. Res.* **221** (1984), 577.
11. Sawicki J. A. and Sawicka B. D., *Hyperfine Interact.* **13** (1983), 199.
12. Philips Data Handbook, Book T9: Electron Tubes, Philips Export B.V., Eindhoven, 1987, p. 333.
13. Sawicki J. A. and Tylisczak T., *Nucl. Instrum. Methods* **216** (1983), 501.
14. Zabinsky J. S. and Tatarchuk B. J., *Nucl. Instrum. Methods Phys. Res., B* **31** (1988), 576.
15. Klingerhöfer G. and Kankeleit E., *Hyperfine Interact.* **57** (1990), 1905.
16. Klingerhöfer G. and Meisel W., *Hyperfine Interact.* **57** (1990), 1911.
17. Murad E., Bowen L. H., Long G. J. and Quin G., *Clay Miner.* **23** (1988), 161.
18. Jamber J. L. and Dutrizac J. E., *Chem. Rev.* **98** (1998), 2549.
19. Vandenberghe R. E. and De Grave E., In: G. J. Long and F. Grandjean, *Mössbauer Spectroscopy Applied to Inorganic Chemistry*, Vol. 3, Plenum Press, New York, 1989, p. 59.
20. da Costa G. M., De Grave E., de Bakker P. M. A. and Vandenberghe R. E. *Clays Clay Miner.* **43** (1995), 656.
21. da Costa G. M., De Grave E. and Vandenberghe R. E. *Hyperfine Interact.* 117 (1998) 207.
22. Morrish A. H., *Canted Antiferromagnetism: Hematite*, World Scientific, Singapore, 1994.
23. De Grave E., Bowen L. H. and Chambaere D., *J. Magn. Magn. Mater.* **30** (1983), 349.
24. Kündig W., Bömmel H., Constabaris G. and Lindquist R. H., *Phys. Rev.* **142** (1966), 327.
25. van der Kraan A. M., *Phys. Status Solidi, A* **18** (1973), 215.
26. Nininger Jr. R. C. and Schroeer D., *J. Phys. Chem. Solids* **39** (1978), 137.
27. De Grave E., Bowen L. H., Vochten R. and Vandenberghe R. E., *J. Magn. Magn. Mater.* **72** (1988), 141.
28. Vandenberghe R. E., Becze-Déak T. and De Grave E., In: I. Ortalli (ed), *Proc. Int. Conf. on Applications of the Mössbauer Effect, Rimini 1995*, SIF, Bologna, 1966, p. 207.
29. Vandenberghe R. E., Van San E., De Grave E. and da Costa G. M., *Czech J. Phys.* **51** (2001), 663.
30. De Grave E., Bowen L. H. and Weed S. B., *J. Magn. Magn. Mater.* **27** (1982), 98.
31. Shinjo T., Kiyama M., Sugita N., Watanabe K. and Takada T., *J. Magn. Magn. Mater.* **35** (1983), 133.
32. Ji-Sen J., Xie-Long Y., Long-Wu C. and Nai-Fu Z., *Appl. Phys. A* **45** (1988), 245.

Hyperfine Interactions (2005) 161:161–169
DOI 10.1007/s10751-005-9178-0 

# Ferrimagnetic to Paramagnetic Transition in Magnetite: Mössbauer *versus* Monte Carlo

J. M. FLOREZ*, J. MAZO-ZULUAGA and J. RESTREPO
*Grupo de Estado Sólido y Grupo de Física y Astrofísica Computacional, Instituto de Física, Universidad de Antioquia, A. A. 1226, Medellín, Colombia; e-mail: jrestre@fisica.udea.edu.co*

**Abstract.** Powder magnetite was analyzed *in situ* via Mössbauer with temperatures ranging from 170 K up to 900 K. Hyperfine fields of the tetrahedral and octahedral sites of magnetite as well as the corresponding average field were followed as a function of temperature in order to elucidate the critical behavior of magnetite at around the Curie temperature. Results evidence a progressive collapse of the Mössbauer spectra onto a singlet-type line at a critical temperature of around 870 K characterized by a critical exponent $\beta = 0.28(2)$ for the hyperfine field. In order to describe such temperature dependence of the hyperfine field, a Monte Carlo-Metropolis simulation based on a stoichiometric magnetite and an Ising model with nearest magnetic neighbor interactions was also carried out. In the model, we have taken into account antiferromagnetic and ferromagnetic interactions depending on the involved ions. A discussion about the critical behavior of magnetite and a comparison between the hyperfine field obtained via Mössbauer and the magnetization obtained via Monte Carlo is finally presented.

**Key Words:** critical exponents, magnetite, Monte Carlo, Mössbauer.

## 1. Introduction

Magnetite has been one of the most studied iron oxides mainly from the experimental point of view [1], and some few simulational works have been published, most of them dealing with micromagnetic calculations in nano-particulate magnetite [2–4]. Nevertheless, to the best of our knowledge, critical exponents of bulk magnetite have not been reported. Additionally, the influence of having different exchange integrals upon the critical behavior remains a subject of much interest [5]. These features have motivated us to consider the Mössbauer and Monte Carlo approaches to elucidate the critical magnetic behavior associated to the ferrimagnetic to paramagnetic transition in magnetite.

## 2. Experimental

Pure magnetite was synthesized by means of a hydrothermal method similar to that described by Cornell & Schwertmann [1]. The sample was prepared by dissolving

* Author for correspondence.

19.88 g of FeCl2 · 4H2O in 200 mL of deionized water flushing with $N_2$. When reaching a temperature of 360 K, a 100 mL mixture containing NaOH 3M and NaNO3 0.3 M is finally added. Zero-field $^{57}$Fe Mössbauer spectra were recorded in standard transmission geometry using a $^{57}$Co source in a Rh matrix. Low temperature measurements were performed by cooling the sample in a closed-cycle helium cryostat with temperatures ranging from 290 K down to 170 K every 20 K. High temperature measurements from 300 K up to 900 K every 50 K where in turn carried out in a Mössbauer furnace at a relative pressure of 0.1333 Pa ($10^{-3}$ Torr). All Mössbauer spectra were fitted by using the Normos program [6] with two sextets ascribed to the tetrahedral and octahedral sites of magnetite.

## 3. Model

As is well established, stoichiometric magnetite crystallizes in an inverse spinel structure with *Fd3m* symmetry having eight $Fe^{3+}$ ions located in tetrahedral sites (A-sites), and eight $Fe^{2+}$ with eight $Fe^{3+}$ ions distributed in octahedral sites (B-sites) per unit cell. Moreover, Mössbauer results on magnetite exhibit a component of mixed valence (2.5+) for octahedral sites attributed to the electron hopping between $Fe_B^{3+}$ and $Fe_B^{2+}$ ions [1]. In the Monte Carlo simulation this fact has been taken into account by means of a flat random distribution of $Fe^{3+}$ and $Fe^{2+}$ ions at B-sites. Magnetic ions $Fe_A^{3+}$, $Fe_B^{3+}$, and $Fe_B^{2+}$ are represented by Ising spins and their interactions are described by the Hamiltonian:

$$H = -\sum_{<i,j>} J_{ij}\varepsilon_i\varepsilon_j\sigma_i\sigma_j. \tag{1}$$

The sum runs over nearest magnetic neighbors or correspondingly next nearest ionic neighbors where oxygen ions are considered as non magnetic. $\sigma_i$ takes on the values ±1 and the $\varepsilon_i$'s are uncorrelated quenched parameters chosen to be 5/2 for $Fe^{3+}$ ions and 2 for $Fe^{2+}$ ions accordingly with the electronic configurations of these ions. Spins interact via antiferromagnetic superexchange interactions when considering $Fe_A^{3+}-Fe_A^{3+}$, $Fe_A^{3+}-Fe_B^{3+}$, $Fe_A^{3+}-Fe_B^{2+}$ couplings whereas ferromagnetic interactions are given for $Fe_B^{3+}-Fe_B^{3+}$, $Fe_B^{3+}-Fe_B^{2+}$, $Fe_B^{2+}-Fe_B^{2+}$ couplings [7]. Such a scenario makes the exchange integral $J_{ij}$ a distributed function of the form:

$$P(J_{ij}) = \sum_{km} x_{km}\delta(J_{ij} - J_{km}), \tag{2}$$

where $x_{km}$ ($k$, $m$ = 1, 2) is the bond density involving a $k$–$m$ coupling with $k$, $m$ = 1 for Fe ions at A-sites, and $k$, $m$ = 2 for Fe ions at B-sites, where in addition $\Sigma x_{km} = 1$ and $\int P(J_{ij})dJ_{ij} = 1$. Particularly, the coefficient $x_{11}$ is the bond density for $Fe_A^{3+}-Fe_A^{3+}$ couplings with superexchange integral $J_{11}$, and so on for other coefficients. The employed numerical values for the integrals are $J_{11} = J_{AA} = -0.42$ meV, $J_{22} = J_{BB} = +2.41$ meV, and $J_{12} = J_{21} = J_{AB} = J_{BA} = -11.15$ meV. These values were fitted in order to reproduce the critical temperature $T_C$ of pure magnetite [7, 8] and the integrals ratio was preserved accordingly with the first principles study of superexchange integrals in magnetite by Uhl and Siberchicot [7] in the nearest neighbor approximation. The choice for Ising spins allows reproducing a case with strong uniaxial anisotropy while keeping computational efforts within reasonable limits. Such a choice is also motivated by the degree of complexity of the system not only by the implicit integrals distribution which are considered in the most realistic way, but by the non-trivial crystalline structure where the existence of two crystallographic sites involves different magnetic coordination numbers, namely $z_{AA} = 4$, $z_{BB} = 6$, $z_{AB} = 12$ and $z_{BA} = 6$ [9]. In the simulation we employ a single-spin flip Metropolis dynamics [10, 11], periodic boundary conditions and several linear system sizes $L$ ranging from 3 up to 15 with a total number of magnetic ions $N = 24 \times L^3$. Simulated annealing from well above the Curie temperature, with an initial random spin configuration, down to 170 K (above the Verwey transition) was carried out. In the vicinity of the critical region the temperature step was as small as 0.2 K. In computing equilibrium averages, around $2 \times 10^4$ Monte Carlo steps per spin (mcs) were considered after equilibration. The main quantities to compute were the magnetization per magnetic site $m = \sum\varepsilon_i\sigma_i/\sum\varepsilon_i$ and the energy $E$ from Equation (1). Specific heat $c$ and magnetic susceptibility $\chi$ were also computed from fluctuation relationships as it has been described elsewhere [10, 11]. Magnetic contributions to the total magnetization per magnetic site arising from every ion sublattice were also analyzed in a differentiated way by considering $m_{Fe^{3+}_A}$ and $m_{Fe_B}$, which allows comparing with the Mössbauer hyperfine fields $B_{hf,A}$ and $B_{hf,B}$, respectively. Programs were implemented in FORTRAN™ using a Pentium IV computer (3.2 GHz, 1 GB RAM). CPU time, depending on system size $L$, was in the range 5–160 h for $L = 3$ and $L = 15$, respectively.

## 4. Results and discussion

Figure 1 shows the Mössbauer spectra for some selected temperatures. As is observed, spectra evolve with temperature from a well-defined sextet pattern toward a single absorption line as the Curie temperature at around 870 K is approached. Such behavior, as expected for a thermally-driven magnetic phase transition, reveals a temperature dependence of the hyperfine fields as that shown in Figure 2, characterized by a sharp drop as the critical temperature $T_C$ is reached. Experimental data close and below $T_C$ were fitted according to the following expression:

$$B_{hf} = \hat{B}_0\left(1 - \frac{T}{T_C}\right)^{\beta}, \tag{3}$$

and it corresponds to the solid lines in Figure 2. Hence, the best fitted values were $T_C = 871 \pm 4$ K and $\beta = 0.28 \pm 0.02$ for the average hyperfine field. Similar dependences for the tetrahedral and octahedral contributions were also obtained

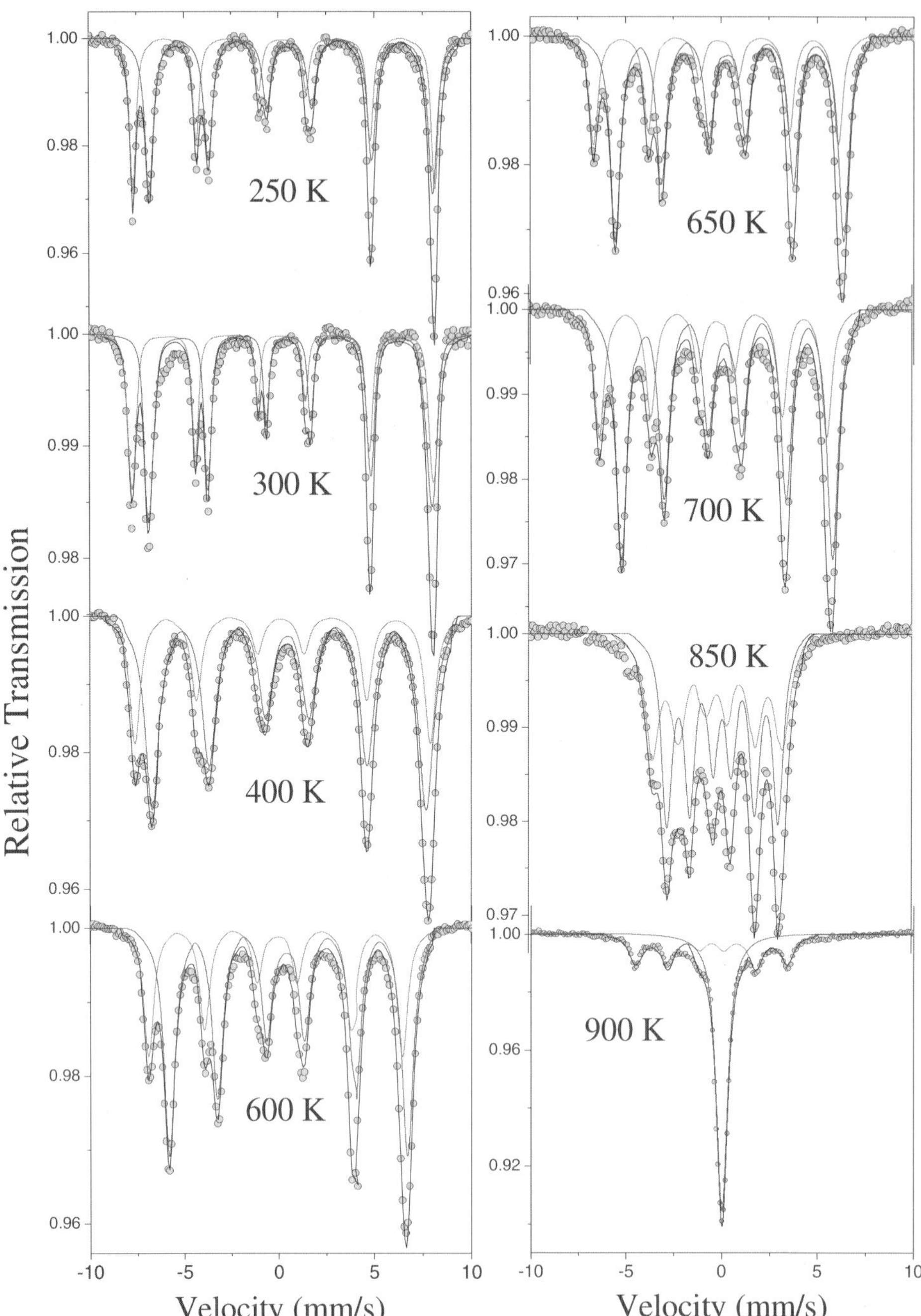


*Figure 1.* Mössbauer spectra for some selected temperatures. Spectra tend to collapse onto a single absorption line as the Curie temperature is approached in agreement with a magnetic phase transition. Magnetic component at 900 K with a hyperfine field greater than those obtained at 850 K is attributed to a partial oxidation of magnetite influenced by the high temperature.

yielding the following values $T_{C,A} = 876 \pm 3$ K, $\beta_A = 0.29 \pm 0.01$, and $T_{C,B} = 865 \pm 4$ K, $\beta_B = 0.27 \pm 0.02$, respectively. The observed differences in the critical exponents can be attributed, in the frame of mean field theory, to the different involved coordination numbers giving rise to different number of bonds per crystallographic site. Additionally, by assuming an occupancy of vacancies equally probable in both sites [12], we obtain an oxidation parameter $x = 0.06$ leading to the formula $Fe_{2.94}O_4$.

Concerning Monte Carlo results, Figure 3 shows the temperature dependence of the total magnetization and their contributions per site for $L = 15$. As is shown in this figure the critical temperature takes place at around 860 K. The occurrence of ferrimagnetic order below $T_C$ is also evidenced by the different magnitudes of the contributions $m_{Fe_A^{3+}}$ and $m_{Fe_B}$ relative to the total magnetization. Such ferrimagnetic order is characterized by an antiparallel alignment between spins of different sublattices with different magnetization values. Figure 4 shows the temperature dependence of the specific heat where a well-defined lambda-type behavior around the Curie temperature ascribed to a ferrimagnetic to paramagnetic transition is observed. Extrapolation to the thermodynamic limit as a function of the linear system size was carried out according to finite size scaling theory [10, 13, 14] by using:

$$T_C(L) \approx T_C(\infty) + aL^{-1/\nu}, \tag{4}$$

where $T_C(L)$ was taken from the maxima of the specific heat and the magnetic susceptibility. The best estimate of the Curie temperature is $T_C(\infty) = 860 \pm 3$ K relatively close to the experimental value ($T_C = 858$ K) [8]. The effective correlation length exponent $\nu = 0.65(2)$ was estimated over the range $|T/T_C - 1| \sim 10^{-3}$ from the maximum values of the fourth order cumulant of the magnetization and the logarithmic derivatives of both the magnetization $\langle |m| \rangle$ and the square of the magnetization $m^2$ following the procedure described by Ferrenberg and Landau [13]. Concerning the $\beta$ exponent, the inset in Figure 4 shows the temperature dependence of the magnetization evaluated at the extrapolated Curie temperature for different $L$ values. From the slope and according to [10, 13, 14]:

$$\langle |m(T_C(\infty))| \rangle \propto L^{-\beta/\nu}, \tag{5}$$

our best estimate for $\beta$ is 0.35(1). This exponent differs slightly from that of the pure Ising model $\beta = 0.3258(44)$ [13]. This fact presumably can be ascribed to the distributed character of the superexchange integrals (see Equation (2)), since as is well established, the critical behavior is rule out by the degree of scattering of the exchange integrals [5].

Finally from Figures 2 and 3, a correlation between the magnetization per magnetic site, including the contributions from $Fe_A^{3+}$ and ($Fe_B^{2+}$, $Fe_B^{3+}$) ions, and the respective hyperfine fields was carried out (see Figure 5). Results reveal a closely linear relationship between the hyperfine field and the magnetization. The different slopes, which are assumed to be proportional to an effective number of

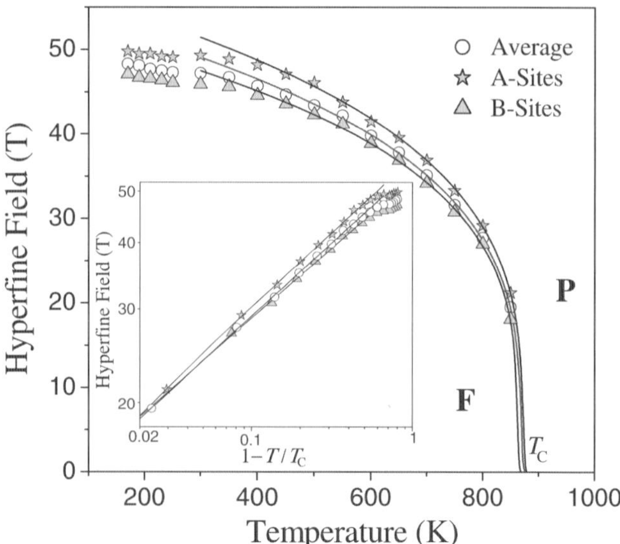


*Figure 2.* Temperature dependence of the hyperfine field obtained via Mössbauer. Contributions of the tetrahedral and octahedral sites are also included. *Inset* shows the same dependence in a semilog plot as a function of $1 - (T/T_C)$. *Solid lines* correspond to the best fitting using Equation (3). Labels *F* and *P* stand for the ferrimagnetic and paramagnetic states, respectively.

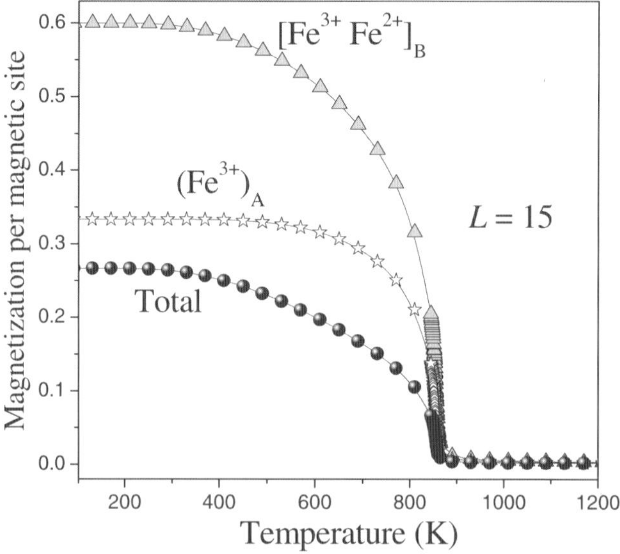


*Figure 3.* Temperature dependence of the total magnetization per magnetic site for $L = 15$ and the respective contributions from $Fe_A^{3+}$ ions and ($Fe_B^{2+}$, $Fe_B^{3+}$) ions. Saturation values of the contributions to the magnetization in the low temperature regime reveal the occurrence of ferrimagnetic order.

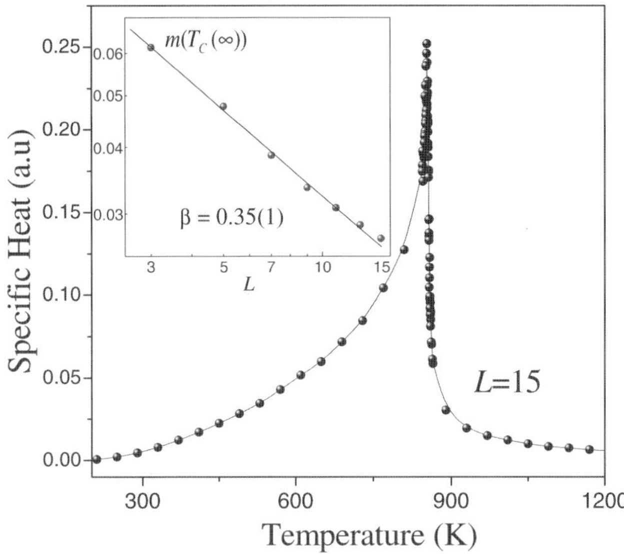


*Figure 4.* Temperature dependence of the specific heat for $L = 15$. Inset shows the log–log plot of the magnetization evaluated at the extrapolated critical temperature to determine the $\beta$ exponent.

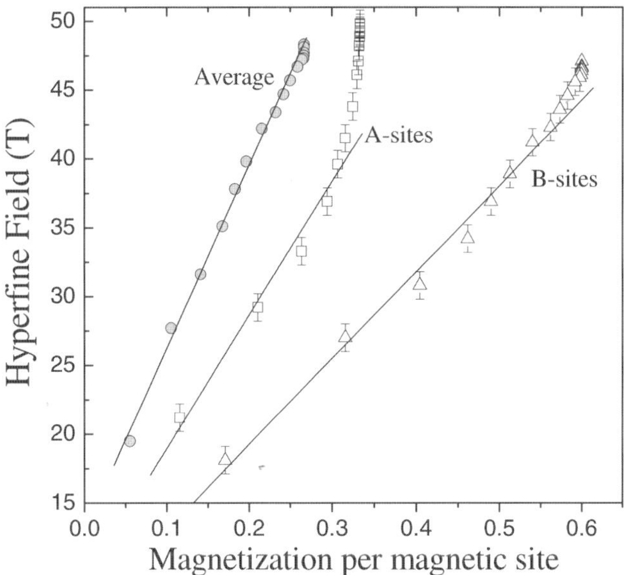


*Figure 5.* Correlation between the hyperfine field and the magnetization per magnetic site inferred from Figures 2 and 3.

magnetic nearest neighbors, are attributed to the different coordination numbers per crystallographic site. In particular, the total number of nearest magnetic neighbors around a $Fe^{3+}_{A}$ ion is 16, whereas that number for a B-ion is 12. On the other hand, the observed linear behavior can be understood in terms of the dipolar contribution to the hyperfine magnetic field acting on the Mössbauer nucleus. In this way, an increase on the magnetization will give rise to a greater magnetic dipolar field and consequently a greater hyperfine field value.

## 5. Conclusions

The critical exponent of the magnetic hyperfine field from Mössbauer measurements in a real magnetite sample, and the exponent $\beta$ of an ideal stoichiometric magnetite in the framework of a three dimensional Ising model, have been computed. Firstly, the difference between the $\beta$ exponent in our Monte Carlo calculation and that of the pure Ising model [13] can presumably be ascribed to the distributed character of the superexchange integrals [5]. Such distribution arises from the different involved integrals as well as from the random distribution of $Fe^{2+}$ and $Fe^{3+}$ ions at B-sites leading to the mixed valence $Fe^{2.5+}$ as evidenced via Mössbauer. Although the $\beta$ exponent for the critical behavior of $B_{hf}$ (0.28) and that for the magnetization via Monte Carlo (0.35) should be similar according to the proportionality on these quantities (Figure 5), the observed difference could be mainly attributed to: i) the non-stoichiometric character of the real magnetite sample contrary to the full stoichiometric character of the simulated single crystal magnetite, ii) the real magnetocrystalline anisotropy different from that of the employed Ising model and iii) the fact of having a low density of experimental points trough the collected Mössbauer spectra in the vicinity of the critical temperature.

Finally, it is important to stress that the employed Monte Carlo approach, provides suitable information on the contributions to the total magnetization of the involved $Fe_A^{3+}$ and ($Fe_B^{2+}$, $Fe_B^{3+}$) sublattices in a differentiated way. This fact allows comparing with the obtained hyperfine fields per crystallographic site through Mössbauer measurements, from which a predominant and closely linear dependence between the magnetization and the hyperfine field was observed.

## Acknowledgements

This work was supported by Universidad de Antioquia-CODI in the frame of the Sostenibilidad Project 2003-2004, Ecos-Nord and the COLCIENCIAS Project 1115-05-12409. One of the authors (J.M.F) would like to acknowledge the financial support provided by Vicerrectoría de Investigación–UdeA with the young researchers program. We are also indebt with Jeaneth Urquijo, who provided the samples for the present work.

## References

1. Cornell R. M. and Schwertmann U., In: *The Iron Oxides*, VCH mbH, Weinheim, Germany, 1996; Halasa N. A., DePasquali G. and Drickamer H. G., *Phys. Rev. B* **10** (1974), 154;

DaCosta G. M., DeGrave E., Bakker P. M. and Vandenberghe R. E., *Clays and Clay Min.* **43**(6), (1995), 656; Evans R. C., In: *An Introduction to Crystal Chemistry*, 2nd Edition, Cambridge University Press, Cambridge, 1966, p.173.

2. Muxworthy A. R. and Williams W., *J. Geophys. Res.* **104** (1999), 29203.
3. Williams W. and Wright T. M., *J. Geophys. Res.* **103** (1998), 30537.
4. Winklhofer M., Fabian K. and Heider F., *J. Geophys. Res.* **102** (1997), 22695.
5. Birgeneau R. J., Cowley R. A., Shirane G., Yoshizawa H., Belanger D. P., King A. R. and Jaccarino V., *Phys. Rev. B* **27** (1983), 6747.
6. Brand R. A., *Nucl. Instrum. Methods Phys. Res. B* **28** (1997), 417.
7. Uhl M. and Siberchicot B., *J. Phys.: Condens. Matter* **7** (1995), 4227.
8. Kittel C., In: *Introduction to Solid State Physics*, 7th Edition, John Wiley & Sons, New York, 1996, p. 449.
9. Iglesias O. and Labarta A., *Phys. Rev. B* **63** (2001), 184416-1.
10. Landau D. P. and Binder K., In: *A Guide to Monte Carlo Simulations in Statistical Physics*, Cambridge University Press, Cambridge, 2000, p. 71.
11. Newman M. E. J. and Barkema G. T., In: *Monte Carlo Methods in Statistical Physics*, Clarendon Press, Oxford, 1999, p. 46.
12. Betancur J. D., Restrepo J., Palacio C. A., Morales A. L., Mazo-Zuluaga J., Fernández J. J., Pérez O., Valderruten J. F. and Bohórquez A., *Hyperfine Interact.* **148/149** (2003), 163.
13. Ferrenberg A. M. and Landau D. P., *Phys. Rev. B* **44** (1991), 5081.
14. Fisher M. E. and Barber M. N., *Phys. Rev. Lett.* **28** (1972), 1516.

Hyperfine Interactions (2005) 161:171–176
DOI 10.1007/s10751-005-9179-z


# Mössbauer Investigation of Fe–Mn–Cu Nanostructured Alloys Obtained by Ball Milling

M. MIZRAHI, S. J. STEWART, A. F. CABRERA and J. DESIMONI*
*Departamento de Física, Facultad de Ciencias Exactas, UNLP. IFLP-CONICET, C C N° 67, 1900 La Plata, Argentina; e-mail: desimoni@fisica.unlp.edu.ar*

**Abstract.** $Fe_{79}Mn_{21}$, $(Fe_{79}Mn_{21})_{90}Cu_{10}$, and $(Fe_{79}Mn_{21})_{80}Cu_{20}$, alloys, prepared by high energy ball milling were characterized by Mössbauer spectroscopy, X-ray diffraction and ac-susceptibility. Results indicate that the Cu addition favors the formation of a FCC phase with two different magnetic states at room temperature, i.e., an antiferromagnetic and a paramagnetic one. Thermal evolution of the Mössbauer spectra revealed the occurrence of a magnetic ordering along a wide temperature range. This behavior is probably related to Fe atoms in FCC-Fe(Mn,Cu) phase having different environments and grain size distribution. Thermal dependence of in-phase ac-susceptibility shows that a long range ordering starts at 240 K for the $Fe_{79}Mn_{21}$ alloy and shifts towards lower temperatures with the Cu content. These results would reflect a long-range magnetic ordering transition with a distribution of ordering temperatures rather than a blocking process of particle single-domains.

## 1. Introduction

Mechanical alloying by high-energy ball milling (HEBM) has been extensively employed to obtain equilibrium and non-equilibrium alloys. During this process, a refinement of crystallite size and the introduction of atomic disorder take place, which alter the physical properties of the as-prepared alloys.

Depending on the Mn concentration, the Fe–Mn system displays $\varepsilon$, $\alpha$ or $\gamma$ phases with HCP, BCC and FCC crystalline structures, respectively. HCP phase is metastable in all composition range, the BCC one is stable at room temperature while the FCC phase is formed at temperatures higher than 500 K. However, the FCC phase can be retained at room temperature by adding low quantity of C, Si, or Cu or by using non-equilibrium processing methods [1]. Particularly, the study of the antiferromagnetic $\gamma$-phase has attracted considerable attention due to its unusual magnetic behavior [1–4]. Recently, some results have been reported about the magnetic behavior of mechanically prepared Fe–Mn–Cu alloys at the Cu-rich region of the phase diagram [5, 6]. However, at low Mn and Cu concentration this ternary system is still few explored.

* Author for correspondence.

In this work, we have employed the ball milling technique to produce metastable FCC FeMn-alloys. To study the effect of Cu addition on the magnetic properties of the Fe-Mn system, we report on structural and magnetic characterization of $Fe_{79}Mn_{21}$, $(Fe_{79}Mn_{21})_{90}Cu_{10}$, and $(Fe_{79}Mn_{21})_{80}Cu_{20}$ alloys. Samples were characterized by Mössbauer spectroscopy at temperatures between 23 and 300 K, X-ray diffraction and ac-susceptibility measurements.

## 2. Experimental

$(Fe_{79}Mn_{21})_{1-x}Cu_x$ alloys, with nominal composition of $x = 0.00$, 0.10, and 0.20, were prepared by mechanical alloying from high purity elemental constituents Fe, Mn and Cu. Ball milling was performed under Ar atmosphere, at approximately 33 Hz, with a Retsch MM2 horizontal vibratory mill using 10 $cm^3$ stainless steel vials with a mass to powder ratio of about 10. All samples were milled for 15 h.

Alloys were characterized by X-ray diffraction (XRD) with a Philips PW1710 diffractometer (Cu-K$\alpha$ radiation). Mössbauer effect (ME) spectra were taken in transmission geometry using a constant acceleration spectrometer with a $^{57}$Co *Rh* source in the 23 to 300 K temperature range. The spectra were analyzed using the NORMOS program [7]. Isomer shifts ($\delta$) are referred to $\alpha$-Fe at RT; ac-susceptibility measurements between 13 and 325 K were carried out using a LakeShore 7130 susceptometer (the field amplitude was 1 Oe, and the frequency varied from 5 to 10 kHz).

## 3. Results and discussion

For all milled samples, the FCC phase is the main component of the XRD patterns (not shown here). The broad diffraction peaks are attributed to the small crystallite size and lattice strain accumulated along the milling process. The most intense peak (111) of FCC phase shows a slight asymmetry on the right side that arises from the BCC (110) line, being more noticeable for the $x = 0.00$ sample. We found that the FCC cell parameter increases with the Cu content from 3.584(1) to 3.615(1) Å, suggesting the incorporation of Cu into the Fe–Mn structure at atomic level. By considering the integral width of the (111) reflection we estimated, by means of the Scherrer's equation, an average grain size of approximately 15 nm for all the samples.

Room temperature Mössbauer spectra show a broad central signal for all the samples. These spectra, recorded at low velocity range scale (Figure 1), can be reproduced using different fitting procedures. One of them assumes a non-magnetic site plus a magnetic unresolved one. The magnetic component can be associated with the FCC phase that, at this Mn concentration, is antiferromagnetic with a Néel temperature of approximately 350 K. The paramagnetic signal comprehends 80% of the relative area and its isomer shift ($\delta$) is similar to that

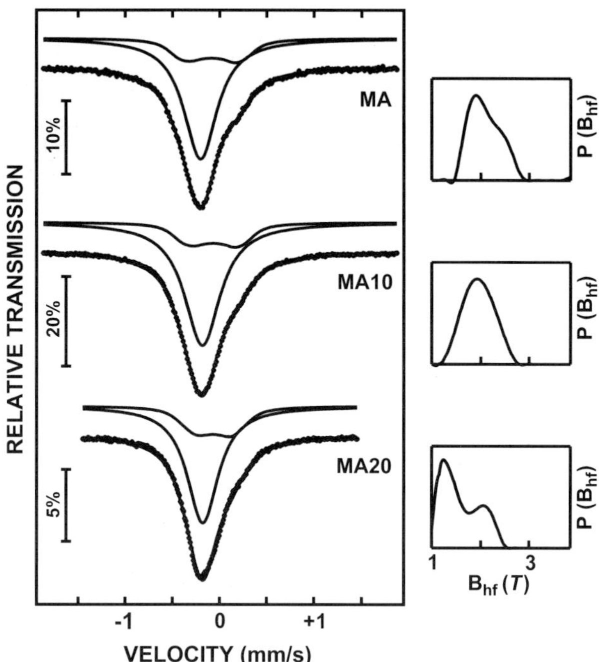


*Figure 1.* Room temperature Mössbauer spectra taken in a low velocity range. The *solid lines* are the result of the fitting procedure assuming a single line and a distribution of hyperfine fields (*left*); and the probability of the hyperfine field distribution (*right*).

reported for the $\varepsilon$-phase [8]. However, the formation of this phase may be discarded because no evidence of it is found by XRD. On the other hand, the $\delta$ values for the non-magnetic signal are different from those of the magnetic FCC site. This, in addition to the lack of a frequency dependence of the ac-susceptibility, would rule out that the single line belongs to a superparamagnetic state of the FCC phase. Furthermore, this signal cannot totally account for iron sites at grain surfaces having a different magnetic state in comparison to the core, because a rough estimation gives that the iron at the surface layer would only take 13% of the relative area. Thus, we assign both components to iron atoms in the $\gamma$-phase having different environments, due to inhomogeneities produced by the milling process, which originates different magnetic states at room temperature. On the other hand, the broad line width of the magnetic site suggests that there is a distribution of hyperfine fields. Therefore, we propose a second fitting procedure consisting of both a distribution of static hyperfine fields and a single line to account for the spectra at RT (Figure 1, Table I). The low $\delta$ values obtained ruled out possible incorporation of Fe into the Cu FCC lattice [9]. On the other side, we observe that the average hyperfine field decreases and the isomer shifts slightly increase with the Cu content. These trends would indicate that the Cu atoms incorporate into the FCC structure.

*Table I.* Hyperfine parameters obtained from the fitting procedure described in the text for the RT spectra taken in a low velocity range

| Sample | $B_{hf}$ (T) | $\delta_d$ (mm/s) | $P$ % | $\delta_x$ (mm/s) |
|---|---|---|---|---|
| $Fe_{79}Mn_{21}$ | 2.21 | 0.02 | 26 | −0.10 |
| $(Fe_{79}Mn_{21})_{90}Cu_{10}$ | 1.95 | 0.04 | 22 | −0.08 |
| $(Fe_{79}Mn_{21})_{80}Cu_{20}$ | 1.60 | 0.05 | 27 | −0.07 |

The errors are considered in the least significant figure. For the percentage of the distribution error is 5%. $B_{hf}$ is the average hyperfine field; $\delta_d$ and $\delta_x$ are the isomer shifts for the distribution and single line, respectively.

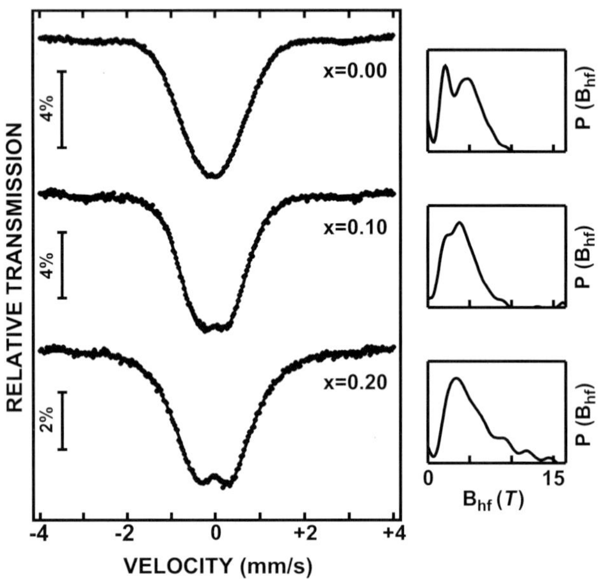


*Figure 2.* Mössbauer spectra recorded at 23 K for $Fe_{79}Mn_{21}$, $(Fe_{79}Mn_{21})_{90}Cu_{10}$, and $(Fe_{79}Mn_{21})_{80}Cu_{20}$, alloys. The *solid lines* are the result of the fitting assuming a distribution of hyperfine fields.

Thermal evolution of the Mössbauer spectra shows a line broadening that starts at temperatures lower than 220 K. This fact suggests that a magnetic ordering or a blocking process is taking place. To simplify the analysis, these spectra were fitted assuming a distribution of hyperfine fields $B_{hf}$ (Figure 2), which also includes a non-vanishing probability at zero fields to account for the paramagnetic component. The resulting average $B_{hf}$, underestimated with respect to the second model above mentioned, increases from its values at RT ($B_{hf} \sim 1-1.3$ T) as the temperature decreases (Figure 3). This increment starts at ~220 K for $Fe_{79}Mn_{21}$ and $(Fe_{79}Mn_{21})_{90}Cu_{10}$, and shifts towards a lower value for the $(Fe_{79}Mn_{21})_{80}Cu_{20}$ alloy (Figure 2). In addition, the average $B_{hf}$ at 23 K is larger for $(Fe_{79}Mn_{21})_{80}Cu_{20}$ (Figure 2). At low temperatures, a small contribution of

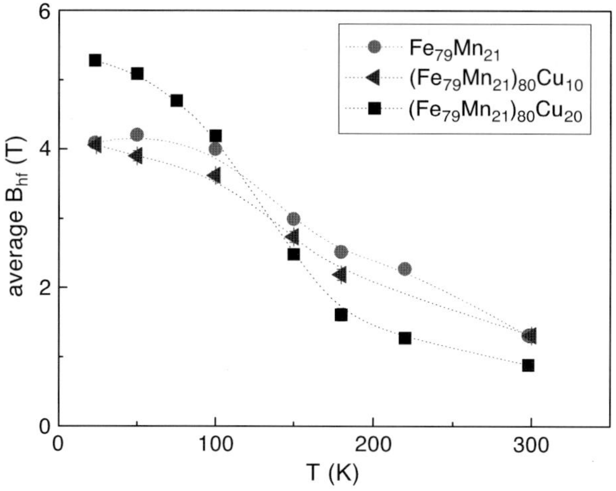


*Figure 3.* Thermal dependence of the average hyperfine field that results from the fitting procedure described in the text.

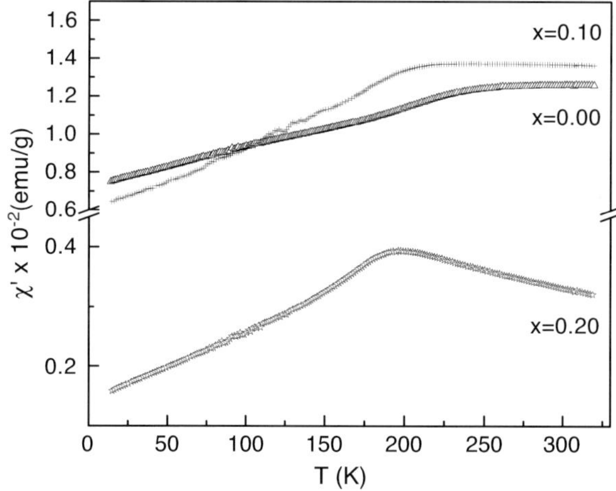


*Figure 4.* Thermal dependence of the in-phase ac-susceptibility for the three samples.

a sextet (inner absorption lines at approximately ± 3 mm/s) associated with the BCC phase was also detected (full spectra are not shown here), in agreement with XRD data. The percentage of this BCC signal decreases with the Cu content.

The in-phase ac-susceptibilities, $\chi'$, show an abrupt change at approximately 240, 215 and 192 K for $x = 0$, 0.10 and 0.20, respectively (Figure 4), in agreement with the temperature at which the Mössbauer line broadening starts. Below those temperatures, $\chi'$ decreases gradually reflecting a long-range magnetic ordering transition with a distribution of ordering temperatures due to the local in-homogeneities in the sample and the distribution of grains sizes. Measurements taken at different frequencies (not shown here) do not present remarkable changes, which reflect the prevalence of a long-range order transition instead of a blocking process of particle magnetic moments.

## 4. Conclusion

Our results show that, by HEBM Fe and Mn (21%) powders stabilize in a FCC $\gamma$-FeMn phase, and a small amount of $\alpha$-Fe(Mn) phase is also present. The BCC contribution decreases with Cu addition, being almost negligible for $x = 0.20$. The FCC phase displays two magnetic states at room temperature, an ordered state whose ordering temperature is above room temperature, and a paramagnetic one that orders below 220 K. These two magnetic states are due to the local in-homogeneities of the sample and distribution of grain sizes. This gives rise to different iron environments having different number of Mn neighbors and, thus, originates a distribution of ordering temperatures. By adding Cu, the same general features remain. The fact that the Cu incorporates to the FCC lattice, evidenced by the resulting structural and hyperfine parameters, originates a decrease of the temperature at which the ordering of the paramagnetic phase starts.

## References


1. Paduani C., Galvao da Silva E. and Pérez Alcázar G. A., *Hyperfine Interact.* **73** (1992), 233.
2. Belozerskiy G., Gittsovich V., Kramar V., Sokolov O. and Khimich Y., *Phys. Met. Metallog.* **35**(3) (1973), 20.
3. Petrov Y., Shafranovsky E., Baldokhin Y. and Kochetov A., *J. App. Phys.* **86**(12) (1999), 7001.
4. Spisak D. and Hafner J., *Phys. Rev., B* **61**(17) (2000), 11569.
5. Restrepo J., Morales A. L., González J. M., Pérez Alcázar G. A., Medina G., Barrero C. A., Tobon J., Pérez G., Arnache O., Betancur J. D. and Giraldo M. A., *Hyperfine Interact.* **134** (2001), 199; Restrepo J. and Greneche J. M., *Hyperfine Interact.* **156/157** (2004), 69; Restrepo J., Greneche J. M. and Gonzalez J. M., *Physica B* **354** (2005), 174.
6. Alocén M. C., Crespo P., Hernando A. and Gonzalez J. M., *J. Non-Cryst. Solids* **287** (2001), 268.
7. Brand R. A., *Nucl. Instrum. Methods Phys. Res., B* **28** (1987), 417.
8. Gauzzi F., Verdini B., Principi G. and Badan B., *J. Mater Sci.* **18** (1983), 3661.
9. Roy M. K. and Verma H. C., *J. Magn. Magn. Mat.* **270** (2004), 186.

Hyperfine Interactions (2005) 161:177–183
DOI 10.1007/s10751-005-9180-6 

# Study of the Morin Transition in Pseudocubic $\alpha$-$Fe_2O_3$ Particles

J. F. BENGOA[1], M. S. MORENO[2], S. G. MARCHETTI[1], R. E. VANDENBERGHE[3] and R. C. MERCADER[4,*]
[1]*CINDECA, Facultad de Ciencias Exactas, UNLP, CICBA, CONICET., Calle 47 No. 257, 1900, La Plata, Argentina*
[2]*Centro Atómico Bariloche, 8400 San Carlos de Bariloche, Argentina*
[3]*Department of Subatomic and Radiation Physics, Ghent University, Proeftuinstraat 86, 9000, Gent, Belgium*
[4]*Departamento de Física, IFLP, Facultad de Ciencias Exactas, Universidad Nacional de La Plata, 1900, La Plata, Argentina; e-mail: mercader@fisica.unlp.edu.ar*

**Abstract.** We have studied the Morin transition in nanostructured pseudocubic $\alpha$-$Fe_2O_3$ particles of about 1.8 μm side. The preparation was carefully chosen to obtain a system with a very narrow crystallite size distribution and particles of homogeneous morphology. Two samples were studied: one without thermal treatment ($\alpha$-$Fe_2O_3$(ap)) and another annealed at 673 K in air for 12 h ($\alpha$-$Fe_2O_3$(an)). Both were characterized by XRD, SEM, TGA and Mössbauer spectroscopy. The results indicate that the Morin transition is suppressed for $\alpha$-$Fe_2O_3$(ap), however, $\alpha$-$Fe_2O_3$(an) has a $T_M \approx 230$ K and the transition is completed over a very narrow temperature range. These results are discussed in connection with the crystallite size, the cell parameters, and the presence of $OH^-$ groups (hydrohematite) or incorporated water (protohematite).

## 1. Introduction

Bulk hematite ($\alpha$-$Fe_2O_3$) undergoes a rotation of its antiferromagnetically coupled magnetic moments at a temperature $T_M \approx 260$ K, known as the Morin transition. Above $T_M$, the magnetic moments lie in the basal plane and are slightly canted bringing about weak-ferromagnetism (WF). Below $T_M$ the magnetic moments are antiparallel along the *c*-axis and the compound is a pure antiferromagnet (AF). The transition is suitably studied by Mössbauer spectroscopy because of the significant change in the quadrupole shift ($2\varepsilon$) from $\cong -0.20$ mm/s for WF to $\cong +0.38$ mm/s for AF [1–3]. $T_M$ can be modified or suppressed by different causes. The presence of impurities or substitution for Fe by different cations [4], the content of water and/or $OH^-$ groups [5, 6] and a small crystallite size [6–8] can alter the Morin transition behavior. The role of the last two points

* Author for correspondence.

is still a matter of discussion. In addition, there are evidences that the rotation of the magnetic moments does not occur gradually, but passes rather by some intermediate states [9].

On the other hand, various techniques have been used to prepare nanocrystallized hematites such as by decomposition of goethite [2] or lepidocrocite [6] or by calcination of ferrric nitrate [1] or sulphate solutions [5]. Moreover, hydrothermal methods enable to obtain hematite consisting of uniform peanut-shaped [10] or pseudocubic-shaped [11] particles.

Toward investigating the effect of the crystallite size and the water or OH content on $T_M$, the present work comprises a preliminary study of the Morin transition behavior in nanostructured pseudocubic $\alpha$-$Fe_2O_3$ particles.

## 2. Experimental

Pseudocubic $\alpha$-$Fe_2O_3$ particles were synthesized by a hydrothermal route following the methodology proposed by Sugimoto *et al.* [11] with minor changes. 100 ml of NaOH solution (5.4 M) were added to 100 ml of well-stirred $FeCl_3$ solution (2.0 M) for 5 min. The stirring was continued during 10 min. The gel was kept at 373 K for 12 days into a tight container. Finally, the product was washed and centrifuged several times, in order to eliminate the NaCl present, and overnight dried in air at 353 K.

Part of the as-prepared compound ($\alpha$-$Fe_2O_3$(ap)) was annealed at 673 K during 12 h in a flow of 150 $cm^3$/min of synthetic and dry air. The sample obtained is denoted ($\alpha$-$Fe_2O_3$(an)).

The materials were characterized by X-ray diffraction (XRD), scanning electron microscopy (SEM), thermal gravimetric analysis (TGA) and Mössbauer spectroscopy. The Mössbauer spectra were recorded with a standard 512 channels spectrometer with transmission geometry. Low temperature measurements were performed in a helium closed-cycle refrigerator at temperatures ranging from 15 to 298 K and at 4 K in a liquid He bath cryostat. A $^{57}$Co in Rh matrix source of nominally 50 mCi was used. Velocity calibration was performed against a 6 μm thick $\alpha$-Fe foil. The spectra were evaluated using a least-squares non-linear computer fitting program with constraints. Lorentzian lines were considered with equal width for each spectrum component.

## 3. Results and discussion

SEM investigation (Figure 1a) reveals that $\alpha$-$Fe_2O_3$(ap) consists of uniform-sized particles of about 1.8 μm side with pseudocubic morphology. XRD shows that the sample is pure $\alpha$-$Fe_2O_3$ with cell parameters $a = 5.030 \pm 0.003$ Å and $c = 13.767 \pm 0.008$ Å . From the line-broadening (Scherrer method) crystallite sizes of $\approx$18 and $\approx$24 nm were derived for the (104) and (110) directions,

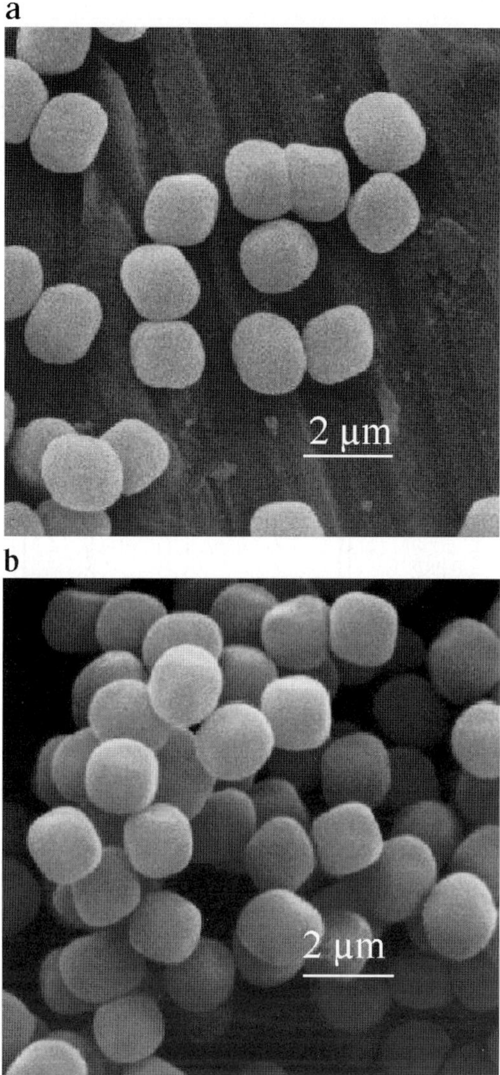


*Figure 1.* (a) SEM of $\alpha$-$Fe_2O_3$(ap) (b) SEM of $\alpha$-$Fe_2O_3$(an).

respectively, showing the pseudocubic particles to be composed of nanosized crystallites. The Mössbauer spectra of $\alpha$-$Fe_2O_3$(ap) did not show any Morin transition down to 22 K. The spectrum at 4K revealed a slight asymmetry indicative of the possible onset of the transition at a very low temperature.

The SEM of the annealed sample $\alpha$-$Fe_2O_3$(an) (Figure 1b) shows that neither the pseudocubic morphology, nor the macroscopic particle size have been modified by the annealing process. X-ray diffraction yields slightly different cell parameters: $a = 5.039 \pm 0.003$ Å and $c = 13.760 \pm 0.008$ Å . Using the same

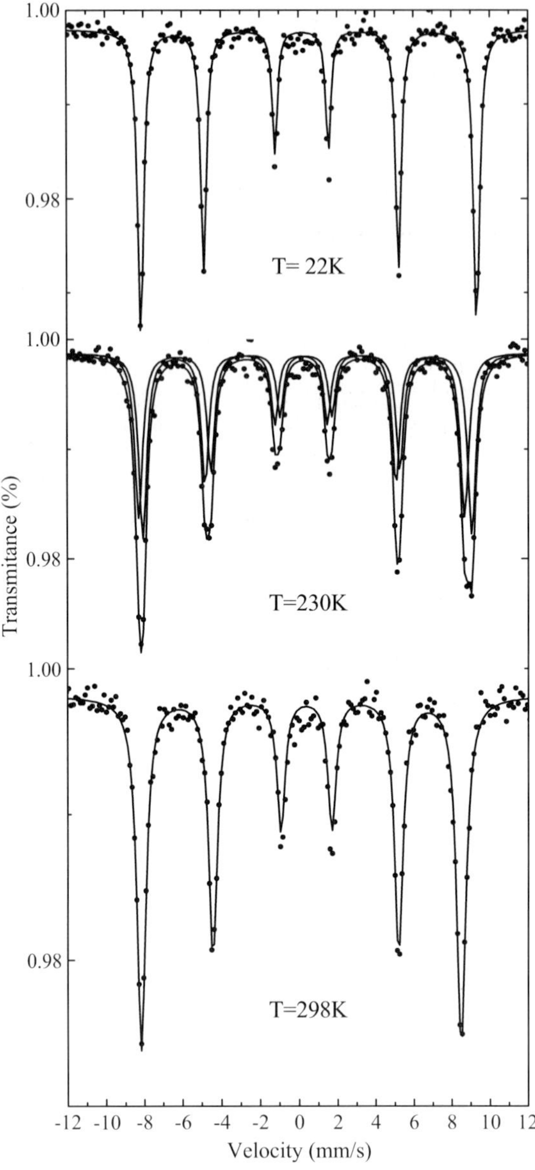


*Figure 2.* Mössbauer spectra of $\alpha$-$Fe_2O_3$(an) at different temperatures.

method as before, the crystallite dimensions turn out to be ≈36 and ≈37 nm in the (104) and (110) directions, respectively. A clear Morin transition at about 230 K is evidenced in the Mössbauer spectra (Figure 2) of $\alpha$-$Fe_2O_3$(an) and by the thermal dependence of its quadrupole shift and hyperfine field (Figures 3 and 4). The transition region $\Delta T_M$ at which both AF and WF coexist spans about 10 K, which is extremely narrow in comparison with other small-crystallite

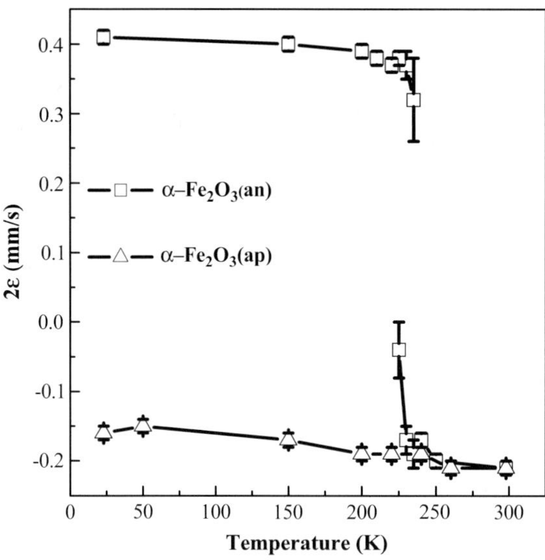


*Figure 3.* Thermal dependence of the quadrupole shift for the indicated samples.

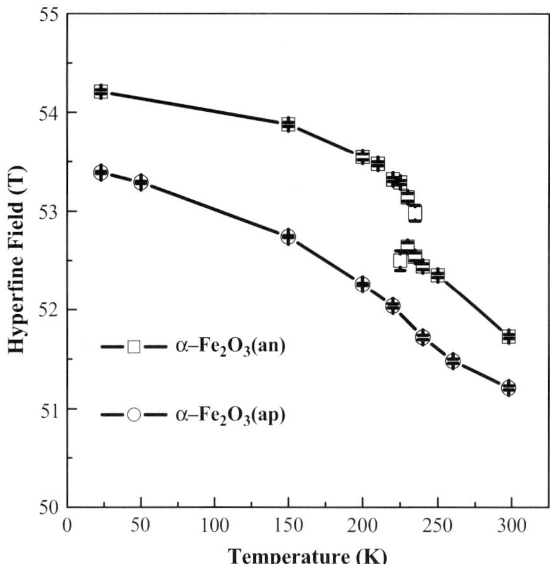


*Figure 4.* Thermal dependence of the hyperfine field for the indicated samples.

hematites of similar size [6, 8]. This evidences that the pseudocubic particles have a very narrow distribution of crystallite sizes.

The strongly different behavior with respect to the Morin transition between the as-prepared and the annealed sample indicates that other causes than the crystallite size, inhibit the Morin transition in the as-prepared sample. Dang *et al.*

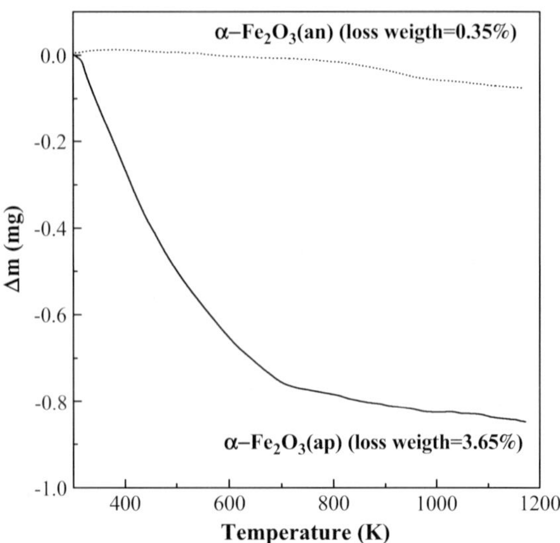


*Figure 5.* TGA for the samples as-prepared and annealed pseudocubic hematite samples.

have shown [5] that the presence of $OH^-$ groups (hydrohematite) and/or incorporated water (protohematite) is responsible for the absence of Morin transition in systems with small – but larger than 20 nm – crystallites. TGA analysis of $\alpha$-$Fe_2O_3$(ap) shows indeed a weight loss of 3.65% of water (Figure 5) whereas the water loss in $\alpha$-$Fe_2O_3$(an) is much less (0.35%). Moreover, the lattice parameters for $\alpha$-$Fe_2O_3$(ap) falls in the *c-a* plot, proposed by Dang *et al.* [5] within the region of hydrohematite for which the Morin transition is suppressed. Otherwise, the lattice parameters for $\alpha$-$Fe_2O_3$(an) are, according to this scheme, more indicative of a (proto)hematite that exhibits the Morin transition.

From a series of lepidocrocite-derived hematites, Vandenberghe *et al.* [8] have shown that $T_M$ decreases in a linear way with the inverse mean crystallite diameter ($1/MCD_{104}$). This line seems to be the upper limit of the Morin transition temperatures of various hydro- and protohematites. For our $\alpha$-$Fe_2O_3$(an) sample with $MCD_{104} = 36$ nm, $T_M$ lies very well on this border line, showing that the small crystallite effect is predominant. On the other hand, the TGA measurements still show some slight decrease above 673 K indicating that there is still some water loss. Hence, we cannot exclude that the effects of some incorporated water (protohematite) and of the crystallite size be closely related.

## 4. Conclusions

It has been shown that the Morin transition can be induced in nanostructured pseudocubic hematite particles by annealing them at moderate temperatures. Such kind of particles are very well suited to investigate the Morin transition behavior because of their uniformity in crystallite sizes. Considering that α-

$Fe_2O_3$(ap) has a crystallite size high enough to experiment a Morin transition and that the thermal treatment did not produce sintering, we can assign the dramatic change in the magnetic behavior to the elimination of water present in the original sample.

## Acknowledgements

The authors acknowledge partial economic support by CONICET (PIP 2853, PIP 2133), ANPCyT (PICT 14-11267), *Comisión de Investigaciones Científicas de la Provincia de Buenos Aires* and *Universidad Nacional de La Plata*, Argentina. This work was conducted within the agreement between FWO-Flanders and SECTIP-Argentina, project FW/PA/02-EIII/003.

## References

1. Nininger R. C. and Schroeer D., *J. Phys. Chem. Solids* **39** (1978), 137.
2. De Grave E., Chambaere D. and Bowen L. H., *J. Magn. Mater.* **30** (1983), 349.
3. Verbeeck A. E., De Grave E. and Vandenberghe R. E., *Hyperfine Interact.* **28** (1986), 639.
4. Tasaki A. and Iida S., *J. Phys. Soc. Japan* **16** (1961), 1697.
5. Dang M. Z., Rancourt D. G., Dutrizac J. E., Lamarche G. and Provencher R., *Hyperfine Interact.* **117** (1998), 271.
6. Van San E., De Grave E., Vandenberghe R. E., Desseyn H. O., Datas L., Barron V. and Rousset A., *Phys. Chem. Miner.* **28** (2001), 488.
7. Gallagher P. K. and Georgy E. M., *Phys. Rev.* **180** (1969), 622.
8. Vandenberghe R. E., Van San E., De Grave E. and da Costa G. M., *Czech. J. Phys.* **51** (2001), 6639.
9. Vandenberghe R. E., Van San E. and De Grave E., *Hyperfine Interact. (C)* **5** (2002), 209.
10. Sugimoto T., Khan M. M., Muramatsu A. and Itoh H., Colloids Surfaces. *A: Physicochem. Eng. Aspects* **79** (1993), 233.
11. Sugimoto T., Sakata K. and Muramatsu A., *J. Colloid Interface Sci.* **159** (1993), 372.

Hyperfine Interactions (2005) 161:185–190
DOI 10.1007/s10751-005-9181-5 

# Anelastic Relaxation Mechanisms Characterization by Mössbauer Spectroscopy

MARTIN JESÚS SOBERÓN MOBARAK
*Secretaría de Educación Pública, México D.F., Mexico; e-mail: msoberon@sep.gob.mx*

**Abstract.** Anelastic behavior of crystalline solids is generated by several microstructural processes. Its experimental study yields valuable information about materials, namely: modulus, dissipation mechanisms and activation enthalpies. However, conventional techniques to evaluate it are complicated, expensive, time consuming and not easily replicated. As a new approach, in this work a Mössbauer spectrum of an iron specimen is obtained with the specimen at repose being its parameters the "base parameters." After that, the same specimen is subjected to an alternated stress–relaxation cycle at frequency $\omega_1$ and a new Mössbauer spectrum is obtained under this excited condition; doing the same at several increasing frequencies $\omega_n$ in order to scan a wide frequencies spectrum. The differences between the Mössbauer parameters obtained at each excitation frequency and the base parameters are plotted against frequency, yielding an "anelastic spectrum" that reveals the different dissipation mechanisms involved, its characteristic frequency and activation energy. Results are in good agreement with the obtained with other techniques.

The study of the mechanical response of a solid subjected to a mechanical influence, at a level sufficiently low to avoid permanent deformation, yields two characteristics of the solid: modulus of elasticity $M$ (or its reciprocal $J$ named compliance modulus) and mechanical energy dissipated.

In ideal elastic deformation the solid absorbs mechanic energy and returns it at recuperation, such way there are not mechanical energy losses in the load–discharge cycle. But in anelastic deformation, part of mechanic deformation energy is used for internal processes such as dislocation migration, interstitials or substitutionals impurities redistribution, etc. In a condition of cyclic charge–discharge work, it conducts to the well-known hysteresis loop where the area of this loop measures the energy dissipated, and the elasticity or compliance modulus becomes complex ($M = M_1 + iM_2$ or $J = J_1 + iJ_2$) with a phase difference angle $\phi$ (Figure 1).

This energy is used at its maximum – internally dissipated – when the characteristic time $\tau$ of the internal process mechanism involved (for example: geometrical kink migration) is reciprocal of load–discharge cycle frequency $\omega$.

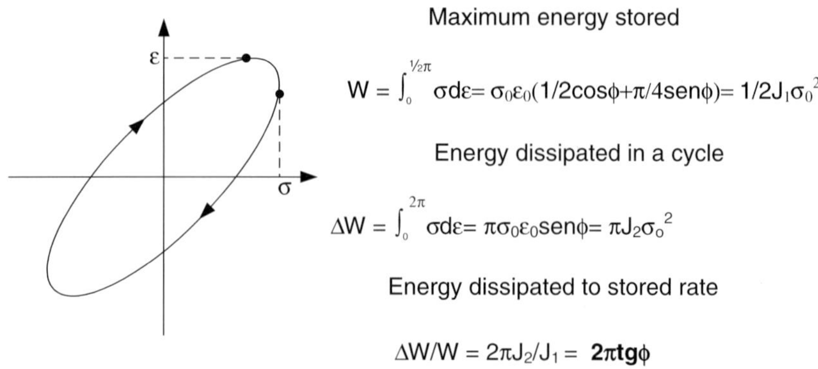


*Figure 1.* Hysteresis loop.

Internally dissipated energy known as internal friction have the following expression:

$$tg\varnothing = (1/\pi)(\Delta W/W)\left[(\omega\tau)/(1+(\omega\tau)^2\right] \tag{1}$$

where $\Delta W/W$ is the ratio of energy dissipated to the maximum stored energy, often called specific damping capacity. The energy dissipation, which is measured by $tg\ \varnothing$, and dynamic (complex) modulus of elasticity $M$ (or its reciprocal $J$ dynamic compliance modulus) are related too:

$$tg\varnothing = (\Delta J/J)\left[(\omega\tau)/(1+(\omega\tau)^2\right] = (\Delta M/M)\left[(\omega\tau)/(1+(\omega\tau)^2\right]. \tag{2}$$

As a function of $\omega\tau$, dynamic modulus of compliance $J$ goes from its fully relaxed value at $\omega\tau = 0$ to its unrelaxed value corresponding at $\omega\tau \rightarrow \infty$ where the stress–strain variations occur so rapidly that no relaxation is possible; finally, maximum occurs when $\omega\tau = 1$.

Anelastic behavior of materials have technical and engineering interest related to rheological flow and fatigue failure. Experimental methods [1, 2] to obtain $tg\ \varnothing$, $\tau$, $J$ or $M$, etc. from an specific solid are very diverse and complex and go from torsion pendulum to ultrasonic attenuation. Generally speaking all these methods are expensive, time consuming and their results depend strongly on sample size, geometrical arrangement, sample bulk and surface treatment and so on; for these reasons their results are hardly reproducible and it has also led to the introduction of different units and ways of presenting the experimental data.

Due to the above mentioned, a new experimental approach was developed and tested successfully based on to evaluate the nuclear response to the anelastic relaxation mechanisms, instead of molecular massive or crystalline response of the solid, using Mössbauer Spectroscopy (MS).

Nuclear response of a material to it's surrounding atmosphere is revealed in the Mössbauer parameters; MS main parameter is recoil-free fraction "*f*," and it is direct function of vibrational state of nucleus in the solid measured by gamma absorption intensity:

$$f = \exp - \left(K_1 + K_2 T^2\right) \tag{3}$$

$K_1 = (3/2)\ (\mathrm{Er}/k\theta)$; $K_2 = (\pi/\theta^2)\ (\mathrm{Er}/k\theta)$; $T$ = solid temperature in °K; Er = gamma recoil energy; $\theta$ = Debye temperature of solid; $k$ = Boltzman constant.

Vibrational state of the nuclii change in response to changes in the surrounding atomic "atmosphere," in such a way that if the solid is unstressed or under static (relaxed) elastic strain, its recoil-free fraction remains the same due to, in both cases, that there are not abnormally active diffusive movement of vacancies, interstitials, substitutionals, dislocations, etc. (static equilibrium has reached); but when the solid suffers a dynamical load–discharge cycle at certain frequency that activate some internal redistribution process, the activated diffusive movement alters the surrounding atmosphere of nuclii changing its vibrational state and, as a consequence, its recoil-free fraction.

So, if a solid not subjected to a mechanical influence has a recoil-free fraction named $f_o$, and the same solid is subjected to a cyclic mechanical influence with frequency $\omega$ that activate some specific internal process, it will shows a new value of recoil-free fraction named $f_\omega$; such change of recoil-free fraction in comparison with $f_o$ is a direct measure of the mechanical energy used for the process or internally dissipated. Then the relative fractional intensity change for several work frequencies becomes:

$$\Delta W/W \alpha \Sigma^i (f_o - f_{\omega i}) \Big/ f_o \alpha\, 2\pi tg\varnothing \tag{4}$$

and internal friction:

$$tg\cdot\varnothing\, \alpha\, \Sigma^i (1/\pi)[1 - (f_{\omega i}/f_o)] \left[(\omega_i \tau) \Big/ \left(1 + (\omega_i \tau)^2\right]\right. \tag{5}$$

and from the standpoint of half-width change of MS peaks, it is possible to determinate the activation energy of each process as:

$$H \alpha \left\{T^2 - C[\ln(\Delta\Gamma/\Gamma)]\right\}^{1/2}. \tag{6}$$

For turning the exposed ideas into experimental method, it is necessary to obtain the Mössbauer spectrum (and its parameters) of a sample under unloaded condition ($f_o$) and under cyclic load–discharge conditions at different frequencies ($f_\omega$); by itself, that reveals changes in the Mössbauer spectrums as shown the next Figure 2.

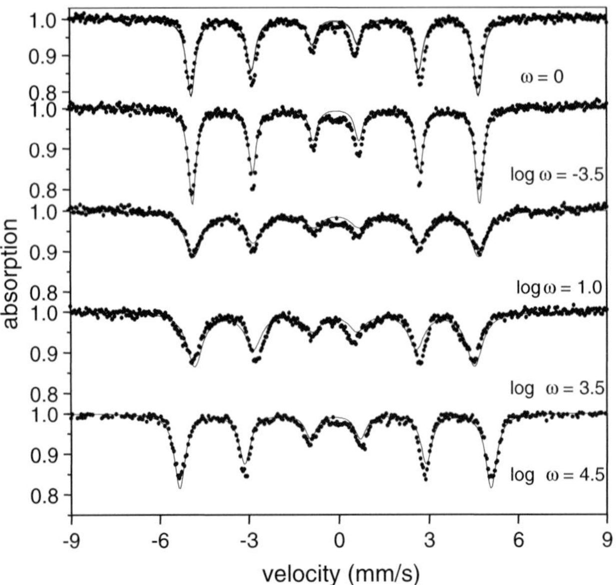


*Figure 2.* Mössbauer spectrum at different frequencies.

Plotting absorption intensity versus log $\omega$ an *anelastic spectrum* is obtained whose peak position yields the characteristic time of the internal friction process revealed by each peak ($\tau = \omega^{-1}$), and peak height yields $\Delta W/W$, tg ∅ and $\Delta J/J$ using expressions (1), (2), (4) and (5) with the adequate constants; in the same way, activation energy of each dissipative mechanism is obtained using expressions (3) and (6) with the different Mössbauer peaks half-width. Plotting also other Mössbauer parameters obtained in each case as: peak width at half maximum, isomer shift, quadrupole split and internal magnetic field, is possible to differentiate and determinate the kind of internal process that generated each peak as shown the Figure 3.

Experimental work was realized using a conventional Mössbauer spectrometer (ASA S-600) with 10 mCi $^{57}$Co source and a exciting device home designed and constructed to provide electromagnetic pulls with variable frequencies from 0 to 1 MHz. Sample used was a foil of interstitials free fully ferritic steel subjected to cyclic stress of E-4 of nominal yield point at room temperature (20 kPa). The sample is sustained by an aluminium frame that fix the specimen for axial stress and prevent lateral vibration. The arrangement is covered with a cage acting as a magnetic shield in order to avoid affect the MS driver as shown the Figure 4. Load–discharge frequency was variated in steps of half order of magnitude due to Debye peaks has width at half maximum of 1.144.

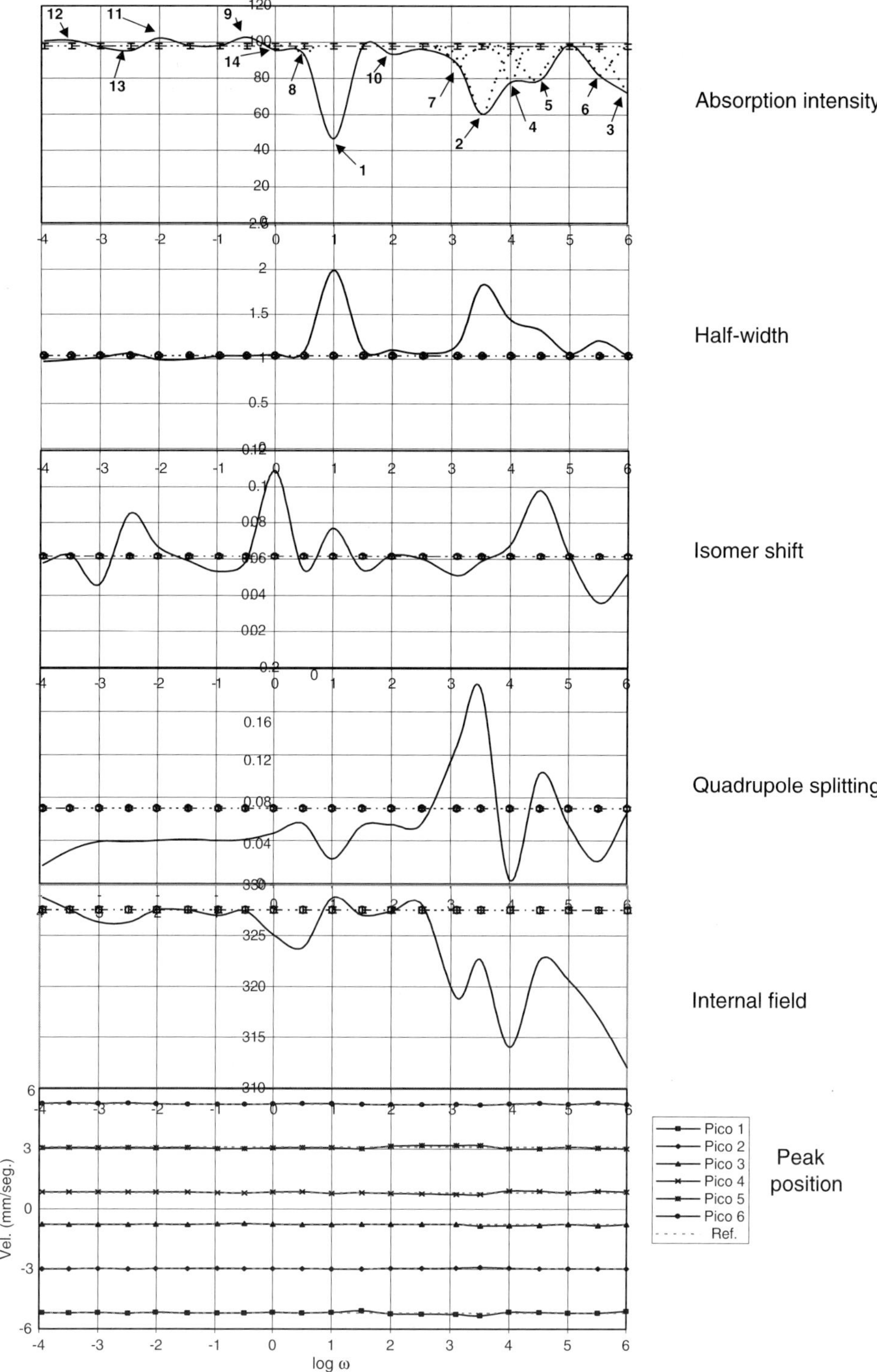


*Figure 3.* Anelastic spectrums obtained (Mössbauer parameter *versus* frequency).

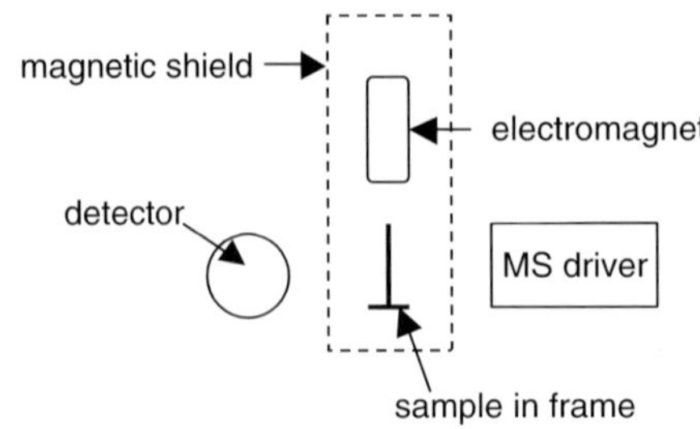


*Figure 4.* Experimental scheme (*lateral view*).

Anelastic spectrum obtained (Figure 3) exhibited 14 peaks and data analysis let identify nine relaxation mechanisms, its characteristic times, its internal friction, its compliance modulus, its relaxation strength and its activation energy (Table I); five of that mechanisms where fully confirmed by other techniques (the rest where no tested and no reports in literature exist) and the only activation energy reported in literature [3] (*H* thermal diff. 0.036 eV) was in excellent agreement with the obtained with this new method (0.0344 eV).

*Table I.* Results

| Peak | Relaxation mechanism | log $\omega$ | $t$ (s) | tg $\phi$ | $\delta J/J$ | $H$ (eV) |
|---|---|---|---|---|---|---|
| 1 | Grain limits | 1 | $10^{-1}$ | 8.33 | 0.1665 | 5.9200 |
| 2 | $\gamma$ Dislocations | 3.51 | $3.09 \times 10^{-4}$ | 6.07 | 0.1213 | 4.2000 |
| 5 | Snoek–Köster H | 4.5 | $3.1 \times 10^{-5}$ | 3.01 | 0.0603 | 1.3000 |
| 6 | Mn diffusion | 5.52 | $3.16 \times 10^{-6}$ | 2.58 | 0.0515 | 0.7200 |
| 14 | H thermal diffusion | 0 | 1 | 0.37 | 0.0074 | 0.0344* |
| *Not confirmed* | | | | | | |
| −3 | Magnetic domains | 6 | $10^{-6}$ | 4.18 | 0.0834 | 0.0263 |
| 7 | $\alpha_1$ dislocations | 3.1 | $7.94 \times 10^{-4}$ | 1.61 | 0.0321 | 0.4600 |
| 8 | Al diffusion | 0.5 | $3.2 \times 10^{-1}$ | 0.80 | 0.0160 | 0.1600 |
| 13 | Cu diffusion | −2.49 | $3.1 \times 10^{2}$ | 0.41 | 0.0081 | 0.0640 |

*Reported value 0.036 eV.

## References

1. DeBatist R., *Mechanics and Mechanisms of Material Damping*; ASTM STP 1169, ASTM, USA, 1992.
2. DeBatist R., Mechanical spectroscopy, In: Cahn R. W., Hassen P., Kramer E. J. (eds.), *Materials Science and Technology*, Vol. 2B. VCH, Germany, 1994, p. 159–217.
3. Heller W. R., *Acta Metall.* **9** (1961), 600.

Hyperfine Interactions (2005) 161:191–195
DOI 10.1007/s10751-005-9182-4


# Comparative Study by MS and XRD of $Fe_{50}Al_{50}$ Alloys Produced by Mechanical Alloying, Using Different Ball Mills

Y. ROJAS MARTÍNEZ[1,*], G. A. PÉREZ ALCÁZAR[2], H. BUSTOS RODRÍGUEZ[1] and D. OYOLA LOZANO[1]
[1]*Department of Physics, University of Tolima, A. A. 546, Ibagué, Colombia; e-mail: yarojas@ut.edu.co; doyolalozano@yahoo.com.mx*
[2]*Department of Physics, University of Valle, A. A. 25360, Cali, Colombia*

**Abstract.** In this work we report a comparative study of the magnetic and structural properties of $Fe_{50}Al_{50}$ alloys produced by mechanical alloying using two different planetary ball mills with the same ball mass to powder mass relation. The $Fe_{50}Al_{50}$ sample milled during 48 h using the Fritsch planetary ball mill pulverisette 5 and balls of 20 mm, presents only a bcc alloy phase with a majority of paramagnetic sites, whereas that sample milled during the same time using the Fritsch planetary ball mill pulverisette 7 with balls of 15 mm, presents a bcc alloy phase with paramagnetic site (doublet) and a majority of ferromagnetic sites which include pure Fe. However for 72 h of milling this sample presents a bcc paramagnetic phase, very similar to that prepared with the first system during 48 h. These results show that the conditions used in the first ball mill equipment make more efficient the milling process.

**Key Words:** mechanical alloying, Mössbauer spectra, X-rays diffraction.

## 1. Introduction

The Fe–Al system is of interest due to its potential commercial application as structural or magnetic materials when they are produced by melting. In the recent years a number of studies were carried on mechanically alloyed FeAl alloys [1–3] milled during 36 h or more, the majority devoted to the magnetic behavior around $x = 0.5$, composition in which the system changes from ferromagnetic to paramagnetic phase. In this work we study samples of the $Fe_{50}Al_{50}$ system prepared by different high energy planetary ball milling and we report a comparative study of the magnetic and structural properties of the alloys.

* Author for correspondence.

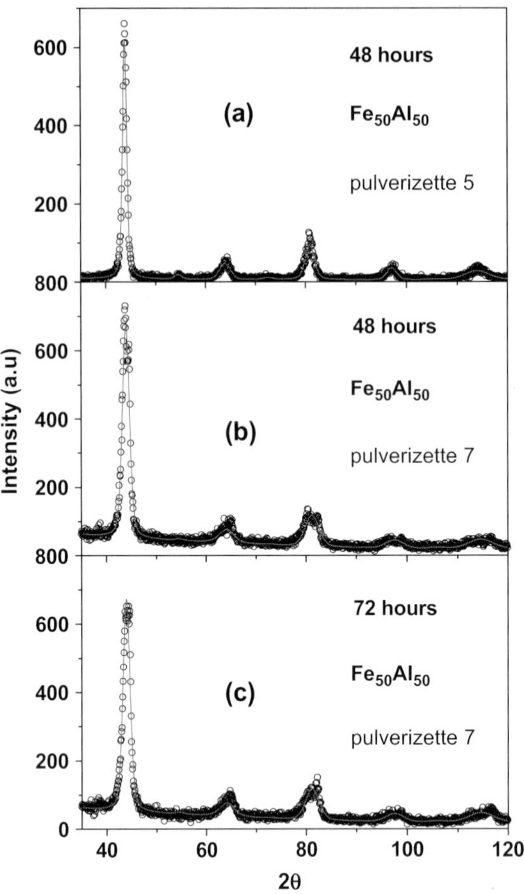


*Figure 1.* X-ray diffraction pattern of $Fe_{50}Al_{50}$ sample using the Fritsch planetary ball mill pulverisette 5 and Fritsch planetary ball mill pulverisette 7.

## 2. Experimental description

The $Fe_{50}Al_{50}$ alloy was prepared by mechanical alloying using high purity iron and aluminum powders in two different planetary ball mills. A Fritsch planetary ball mill pulverisette 5 of high energy with hardened stainless steel vials 500 ml of volume and balls of the same material with 20 mm of diameter and another Fritsch planetary ball mill pulverisette 7 of high energy with hardened stainless steel vials 50 ml of volume and balls of the same material with 15 mm of diameter. The speed was of 280 rpm and the ball to powder mass ratio used was 16:1 in both cases. The milling times were 48 and 72 h. The powders of the milled samples were measured in a transmission Mössbauer spectrometer using a radioactive Co-57/Rh source. The spectra were fitted with hyperfine field distribution (HFD) and doublets using the MOSFIT program [4]. The α-Fe

*Table I.* Results of the XRD parameters of the samples of the $Fe_{50}Al_{50}$ system using Fritsch planetary ball mill pulverisette 5 and Fritsch planetary ball mill pulverisette 7

| Ball mill pulverisette 5 | | Ball mill pulverisette 7 | |
|---|---|---|---|
| | 48 h | 48 h | 72 h |
| a (Å) | 2.908 ± 0.001 | 2.906 ± 0.001 | 2.891 ± 0.002 |
| Mean grain size (nm) | 16.053 ± 0.489 | 14.022 ± 1.491 | 13.621 ± 0.934 |

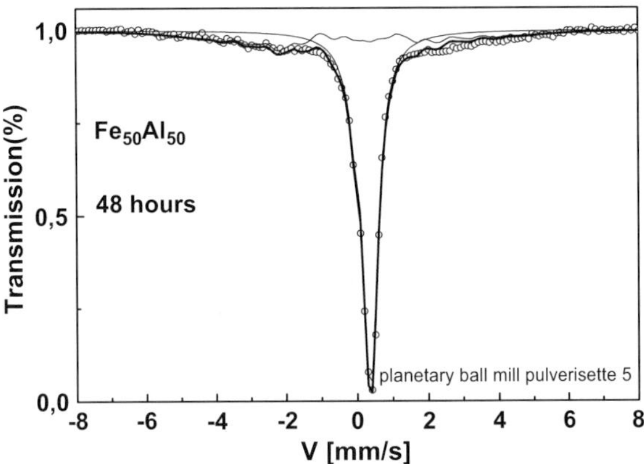


*Figure 2.* Mössbauer spectrum of $Fe_{50}Al_{50}$ alloy with 48 h of milling using the Fritsch planetary ball mill pulverisette 5.

patter was used as calibration sample. The X-ray analysis to establish the structure of the lattice were performed at room temperature for all samples using a RINT2000l diffractometer with the Cu K$\alpha$ radiation and the patterns were fitted using Maud program [5].

## 3. Results and discussion

Figure 1 shows the X-rays diffraction patterns of samples milled during 48 h in the Fritsch planetary ball mill pulverisette 5 (a), and that of the samples milled during 48 (b) and 72 (c) h in Fritsch planetary ball mill pulverisette 7. These patterns were fitted by using the Maud program [5], with which you determines the lattice parameter and the mean grain size. All the patterns indicate that the samples are in the phase bcc, corresponding to the FeAl alloy. The lines of the patters (b) and (c) are broader than that of the pattern (a).

The lattice parameter is nearly the same (~2.90 Å) for samples milled during 48 h and a little small for the sample milled during 72 h. The mean grain size of sample (a) is 16.053 ± 0.489 nm and those of the samples (b) and (c) are 14.022 ± 1.491 and 13.621 ± 0.934 nm, respectively. This can explain the broadening of

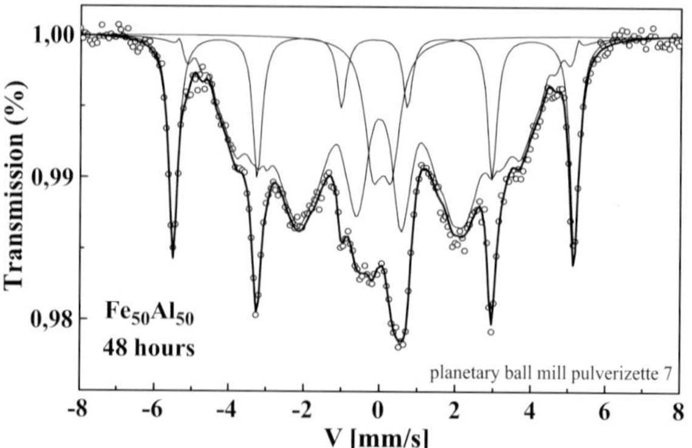


*Figure 3.* Mössbauer spectra of $Fe_{50}Al_{50}$ alloy with 48 h of milling using the Fritsch planetary ball mill pulverisette 7.

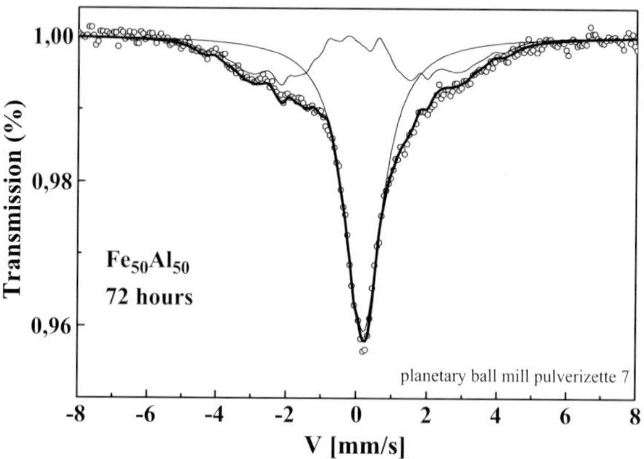


*Figure 4.* Mössbauer spectra of $Fe_{50}Al_{50}$ alloy with 72 h of milling using the Fritsch planetary ball mill pulverisette 7.

the lines observed in patterns (b) and (c). Table I report the results obtained from the fits of the patterns.

Room temperature (RT) Mössbauer spectra for $Fe_{50}Al_{50}$ using the Fritsch planetary ball mill pulverisette 5 with milling time of 48 h, is shown in Figure 2. The fit was performed with a hyperfine field distribution (HFD) and a wide single paramagnetic site ($\Gamma = 0.52$ mm/s), as a result of the greater Al content, which behaves as a magnetic hole. This result proves that the obtained bcc phase is disordered with ferromagnetic and paramagnetic sites with a symmetric charge distribution.

Figure 3 shows the room temperature (RT) Mössbauer spectrum for $Fe_{50}Al_{50}$ using the Fritsch planetary ball mill pulverisette 7 with milling time of 48 h. The spectrum was fitted with a broad paramagnetic site (doublet), a sextet of $\alpha$-Fe (32.93 T) and a hyperfine field distribution. Then, it can be observed that in this case the bcc FeAl phase is not totally consolidated due the pure Fe is detected. In the present case the paramagnetic line of the bcc phase appears as a doublet showing the asymmetry of the charge distribution.

However, for 72 h of milling using the Fritsch planetary ball mill pulverisette 7, the $Fe_{50}Al_{50}$ alloy presents a very similar spectrum to that showed for the sample prepared with the Fritsch planetary ball mill pulverisette 5 during 48 (Figure 2), as it is shown in the Figure 4. The spectrum was also fitted with a broad paramagnetic single line ($\Gamma = 1.22$ mm/s) and a HFD. Comparing the wide of the single paramagnetic line of the first sample with that of the present one it can be note that the latest is much bigger. Then it can be conclude that the alloy consolidation in the present case is still in process.

## 4. Conclusion

Mechanical alloying has been used to produce $Fe_{50}Al_{50}$ alloys. The obtained alloy for 48 h using the Fritsch planetary ball mill pulverisette 5 is totally consolidate in the bcc phase and behaves as a ferromagnetic disordered system with a tendency to a paramagnetic behavior, while those obtained for 48 and 72 h using the Fritsch planetary ball mill pulverisette 7 are in the consolidation process of the bcc phase.

With these results it can be also conclude that the ball mill pulverisette 5 is more efficient to alloy due that its jars are bigger and the balls can give more mean kinetic energy to the powder in each impact.

## Acknowledgements

We are grateful to the Central Committee of Research of the University of Tolima for its financial support and to the Mössbauer Laboratory of the University of Valle for the achievement of this work.

## References

1. Pekala M. and Oleszak D., *J. Magn. Magn. Mater* **157** (1996), 231.
2. Eelman D. A., Dahn J. R., Mackayu G. R. and Dunlap R. A., *J. Alloys and Compounds* **266** (1998), 234.
3. Pochet P., Tominez E., Chaffron L. and Martin G., *Phys. Rev. B* **52** (1995), 4006.
4. Bremers H., Jarms C., Hesse J., Chajwasiliou S., Efthiadis K. G. and Tsonkalas Y., *J. Magn. Magn. Mater.* **140** (1995), 63.
5. Lutterotti L. and Scardi P. J., *Appl. Crystallogr.* **23** (1990), 246.

Hyperfine Interactions (2005) 161:197–202
DOI 10.1007/s10751-005-9183-3 

# First Principles Determination of Hyperfine Parameters on *fcc*-$Fe_8X$ (X = C, N) Arrangements

E. L. PELTZER Y BLANCÁ[1], J. DESIMONI[2] and N. E. CHRISTENSEN[3]
[1]*Grupo de Estudio de Materiales y Dispositivos Electrónicos (GEMyDE), Facultad de Ingeniería, UNLP, IFLYSIB-CONICET C.C. N°565, 1900 La Plata, Argentina*
[2]*Departamento de Física, Facultad de Ciencias Exactas, UNLP, IFLP-CONICET, C.C. N°67, 1900 La Plata, Argentina*
[3]*Department of Physics and Astronomy, Aarhus University, Ny Munkegade, Bld. 520, DK-8000 Aarhus C, Denmark*

**Abstract.** In order to investigate some of the fundamental physical properties of the *fcc*-FeX (X = C, N) austenite solid solutions, we compare the hyperfine parameters obtained by Mössbauer $^{57}$Fe spectrometry and those obtained by the full-potential linear augmented-plane wave (FLAPW) method. We have focused the study on isomer shifts and quadrupole splittings at Fe sites obtained by FLAPW assuming an $Fe_8X$ structure to sketch the austenite. In the present work, we will discuss this point and compare the results of the calculations with experimental data.

**Key Words:** *ab-initio* calculations, austenite, hyperfine parameters.

## 1. Introduction

Steels and cast irons are very useful for different industrial applications like aggressive environment condition devices, high-density recorders, molds and automotive pieces. The present and the future applications of these materials require knowledge of the basic physical properties. Impurities like C or N introduced in the matrix of iron in the face centered cubic (*fcc*) structure produce very important changes in the physical properties, in particular as results of changes in the electron distribution around Fe atoms. Mössbauer spectroscopy [1–4] has successfully contributed in the determination of the hyperfine parameters and of the interstitial atom distribution in austenite. Several models for the arrangement of the solute atoms in the *fcc*-Fe lattice are found in literature [2, 3]. These models are based on two sublattices, an *fcc* sublattice for the Fe atoms and one for the mixture of solute atoms and vacant octahedral interstices, that differ in the probability of occupation of the interstices [2, 3]. Structures of the type $Fe_8X$ and $Fe_4X$, with the solute atoms at the center of the cube of the *fcc* lattice, have been considered. The Full-potential Linearized Augmented-Plane Wave (LAPW) method [5] has been previously applied to calculate the electronic

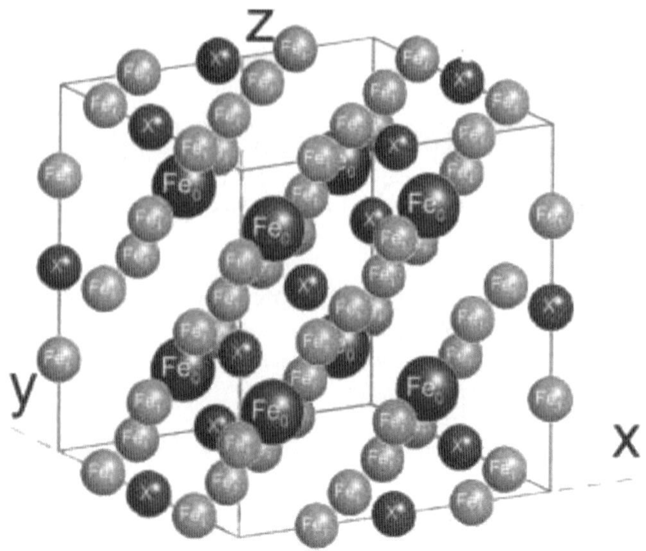


*Figure 1.* $Fe_8X$ cell.

properties, which are intimately correlated with the solute distribution and the crystal structure. The $Fe_8X$-type structure was suggested to reproduce the Fe-C system, while the $Fe_4X$-type structure simulated better the Fe–N austenite [6–8]. A systematic comparison of the hyperfine parameters obtained by Mössbauer spectroscopy on austenite [1–4] and FLAPW calculations has not yet been attempted. The magnetic properties of interstitial solutions are still a matter of discussion.

We present here *ab-initio* calculations of the density of states; magnetic moments, isomer shifts and quadrupole splittings at the iron sites in *fcc*-$Fe_8X$ (X = C, N) arrangement. We report preliminary calculations of isomer shifts ($\delta$) and electric field gradient (EFG) obtained by FLAPW assuming an $Fe_8X$-type structure for the austenite. We discuss the results of the calculations and compare them with Mössbauer data [1–4].

## 2. Calculations

Figure 1 shows the $Fe_8X$ crystal cell chosen for the calculations, with the space group Fm-3m (225) and the lattice parameter depending on which solute atom is present in the structure, if N is in it, $a$ = 13.849 a.u. [9], while if the atom is C, $a$ = 13.839 [9]. Two different Fe environments are distinguished: Fe atoms with one solute atoms in the first interstitial shell ($Fe_1$), and Fe atoms without nearest neighbour X atoms ($Fe_0$).

*Table I.* Theoretical and experimental lattice parameters and hyperfine parameters

| | $Fe_8N$ | | $Fe_8C$ | |
|---|---|---|---|---|
| | Theoretical | Experimental | Theoretical | Experimental |
| a (a.u.) | 13.52 | 13.85[9] | 13.52 | 13.84[9] |
| $B_0$ (Mbar) | 2.83 | | 2.77 | |
| $\delta_0$ (mm/s) ($Fe_0$) | 0.02 | 0.04[2] | | 0.03[2] |
| | | −0.11[2] | −0.04 | −0.10[2] |
| | | −0.08[3] | | −0.05[3] |
| $\delta_1$ (mm/s) ($Fe_1$) | −0.01 | 0.10[2] | | 0.00[2] |
| | | 0.20[2] | −0.01 | 0.01[3] |
| | | 0.04[3] | | |
| EFG ($10^{21}$ V/$m^2$) ($Fe_1$) | −1.75 | 3.1[2] | | 3.1[2] |
| | | 1.69[2] | −4.15 | 2.88[3] |
| | | 1.51[3] | | |

The $\delta$ values are referred to α-Fe at RT.

The *ab-initio* calculations were performed by using a local density approximation (LDA) to the density-functional theory, but general gradient corrections (GGA) were included. All electrons are treated self-consistently, the core fully relativistically, and the valence electrons, scalar relativistically (spin-orbit splitting omitted). In order to improve upon linearization and treat the semicore and valence states consistently in one energy window [10], local orbitals (LO) were included in the basis. Brillouin zone integrations were performed using the tetrahedron method with meshes sufficiently dense to achieve total energy convergence better than $10^{-6}$ Ry. The muffin tin radii were chosen to be (in atomic units) 1.9 for Fe and 1.5 for C or N. The number of k-points was 84 in the irreducible Brillouin zone. In the present work $Fe_8X$ was treated as a non-magnetic compound.

## 3. Results and discussion

The lattice values corresponding to the minimum of energy obtained from the FLAPW calculations are close (≅ 2%) to the experimental ones (see Table I).

The DOS of *fcc*-Fe is affected by the presence of the C- or N-atoms, the p-orbitals of C and N lie between 0 and approximately −9 eV (see Figure 2). In this same range, the Fe atoms also contribute to the DOS. The main contribution to the density from the s-orbitals of the C and N are below of the −10 eV and −15 eV, respectively.

In Table I, the calculations of δ and EFG are compared with the average values recorded at room temperature for *fcc*-FeX alloys with different solute atomic fraction $y$ [1–4] ($0.052 < y_C < 0.085$ and $0.102 < y_N < 0.111$ for Fe–C and Fe–N austenites, respectively). The theoretical calculations associated to the $F_0$

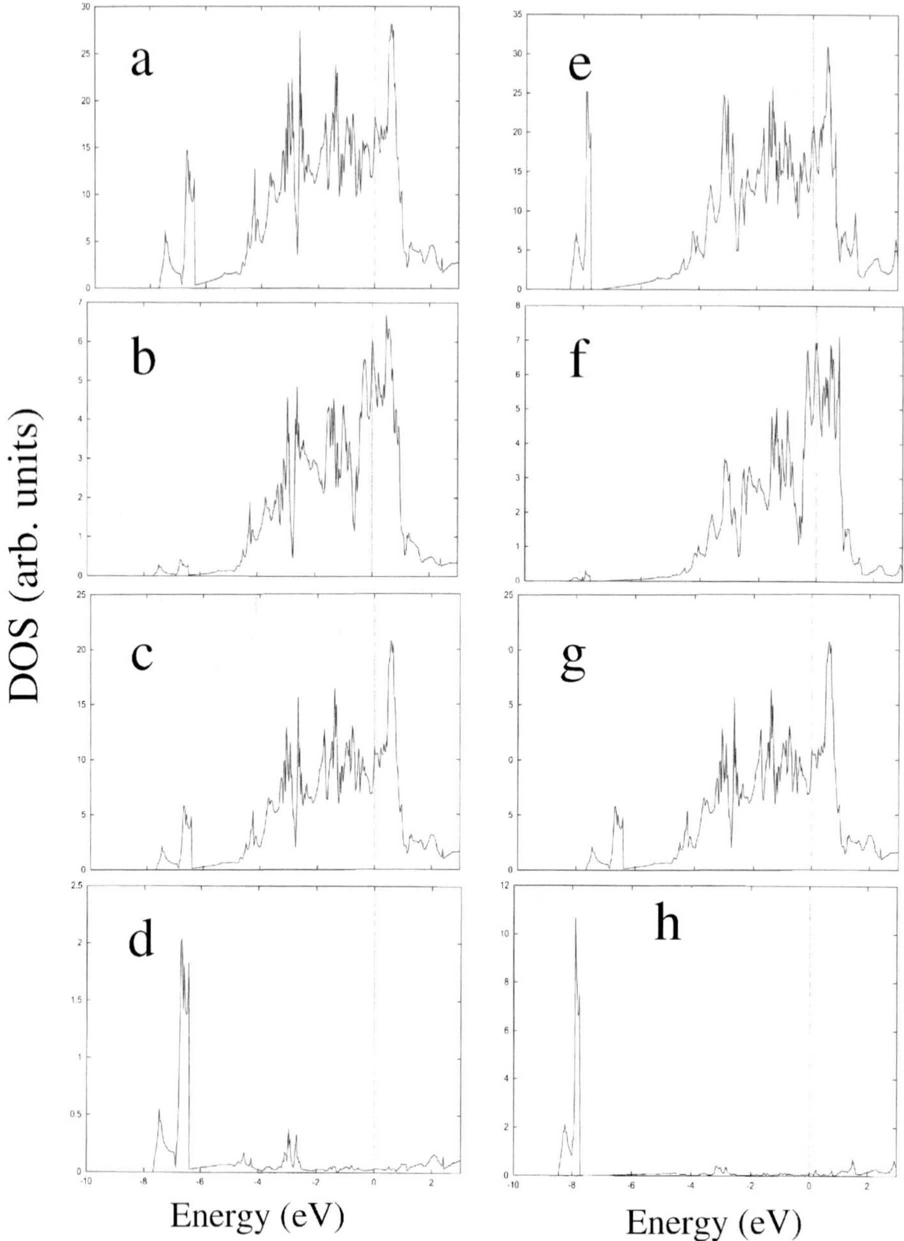


*Figure 2.* DOS calculations. On the *left* $Fe_8C$: (a) Total DOS, (b) $Fe_0$ DOS, (c) $Fe_1$ DOS, and (d) C DOS. On the *right* $Fe_8N$: (e) Total DOS, (f) $Fe_0$ DOS, (g) $Fe_1$ DOS, and (h) N DOS.

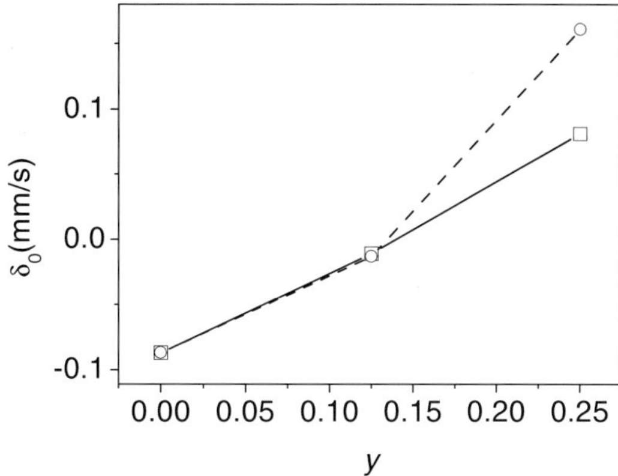


*Figure 3.* Calculated isomer shift for different solute atomic fractions *y*. *Squares*: $Fe_8C$ and circles $Fe_8N$.

site are matched up to the experimental ones ascribed to Fe atoms without solute atoms in the first interstitial shell and with a variable number of solute atoms in the second interstitial shell [1–4]. In the case of site $Fe_1$, the theoretical values are contrasted with the hyperfine parameters belonging to Fe environments with one solute atom in the first independently of the number of solute atoms in the second interstitial shell [1–4].

The experimental isomer shift shows a spread of values in a wide range [1–4]. The theoretical isomer shifts associated to both iron sites ($Fe_0$ and $Fe_1$) seems to be smaller for the Fe–C than the Fe–N solid solution, suggesting the decrease of the density of s-electrons at the iron site in the Fe–N system that is coherent with the calculated density of states. In Figure 3, the present $Fe_0$ site isomer shift calculated values are compare with those obtained under similar calculation conditions for *fcc*-Fe [11] and for the $Fe_4X$ structure [8], indicating that isomer shift depends on the solute concentration. The electric field gradient values of $Fe_1$ site are smaller for the Fe–N than for the Fe–C case, suggesting that the perturbation of the electronic structure of an iron atom by the presence of a neighbouring interstitial is greater for C than for N. Since in the case of polycrystalline samples, Mössbauer spectroscopy only allows us the determination of the absolute values of the EFG. Consequently, the modulus of theoretical should be compared with the experimental results. In this frame, and due to the scatter in the experimental results, the calculated values and the experimental ones are of the same order as observed in Table I.

The $Fe_8X$ crystalline structure proposed in the present work does not represent satisfactory the austenite as a non-stoichiometric compound ($Fe_8X_{1-z}$) as suggested in Refs. [6, 7]. This can be related to the strong dependence of the theoretical parameter on the solute concentration.

## 4. Conclusions

The calculations done in the present work seem to indicate that the hyperfine parameters are strongly dependent on the solute concentrations. The choice of $Fe_8X$ cell does not represent the distribution of solute atoms in austenite.

## References

1. Desimoni J., *Hyperfine Interact.* **156** (2004) 505.
2. Uwakweh O. N. C., Bauer J. P. and Genin J. M., *Metall. Trans.,* **A21** (1990) 589.
3. Oda K., Fujimura H. and Ino H., *J. Phys., Condens. Matter* **6** (1994) 679.
4. Laneri K., Desimoni J., Zarragoicoechea G. J. and Fernández Guillermet A., *Phys. Rev.,* **B66** (2002) 134201.
5. Blaha P., Schwarz K. and Luitz J., WIEN97: *A Full Potential Linearized Augmented Plane Waves Package for Calculating Crystal Properties*, Vienna University of Technology. Getreidemarkt 9/158. A-1060 Vienna, Austria, 1997.
6. Timoshevskii A. N., Timoshevskii V. A. and Yanchitsky B. Z., *J. Phys., Condens. Matter* **13** (2001) 1051.
7. Timoshevskii A. N., Timoshevskii V. A., Yanchitsky B. Z. and Yavna V. A., *Comput. Mater. Sci.* **22** (2001) 99.
8. Peltzer y Blancá E. L., Desimoni J. and Christensen N. E., *Physica B* **354** (2004) 341.
9. Cheng L., Bötteger A., De Keijser T. H. and Mittemeijer E. J., *Scr. Metall. Mater.* **24** (1990) 509.
10. Singh D., *Phys. Rev.,* B **43** (1994) 6388.
11. Peltzer y Blancá E. L., Rodriguez C. O., Shitu J. and Novikov D. L., *J. Phys., Condens. Matter* **13** (2001) 9463.

Hyperfine Interactions (2005) 161:203–209
DOI 10.1007/s10751-005-9184-2


# Magnetic and Structural Properties of the Mechanically Alloyed $Nd_2(Fe_{100-x}Nb_x)_{14}B$ System

D. OYOLA LOZANO[1,*], L. E. ZAMORA[2], G. A. PEREZ ALCAZAR[2], Y. A. ROJAS[1], H. BUSTOS[1] and J. M. GRENECHE[3]
[1]*Department of Physics, University of Tolima, A.A.546 Ibagué, Colombia; e-mail: doyola@ut.edu.co*
[2]*Department of Physics, University of Valle, A.A.25360 Cali, Colombia*
[3]*Laboratoire de Physique de l'Etat Condensé, UMR CNRS 6087, F72085 Le Mans Cedex 9, France*

**Abstract.** In this work we report the magnetic and structural properties obtained by Mössbauer spectrometry, Vibrating Sample Magnetometer and X-ray diffraction of milled powders with initial composition $Nd_2(Fe_{100-x}Nb_x)_{14}B$ with $x = 0$ and $x = 4$. The mixtures were ball milled for different times up to 240 h. Structural and microstructural parameters were derived from high statistics X-ray patterns and discussed as a function of milling time. The Mössbauer spectra of the samples were fitted by means of a sextet and an hyperfine field distribution, associated to a pure iron phase ($\alpha$-Fe) and a disordered iron-based phase, respectively. The $\alpha$-Fe grain size decreases from 50 nm for 6 h up to 5 nm for 240 h milling time. The Vibrating Sample Magnetometer results allow to conclude that these samples behave as soft ferromagnets.

## 1. Introduction

Mechanical alloying is one of the most efficient methods to produce nano-structured powders. When the solid state reaction is achieved by using this technique, it becomes thus advantageous, because melting and solidification are bypassed. The discovery in 1983 of Nd/Pr–Fe–B system [1, 2] has opened a new research field for novel permanent magnets, with relatively high coercive field, high remanence and high magnetic energy product. A possible method to improve the magnetic properties in permanent magnets is to use nanocomposites consisting of a mixture of hard and soft magnetic phases, where the soft phase contributes to a high saturation magnetization while the hard phase originates a

* Author for correspondence.

high coercive field. Kneller and Hawig [3] concluded that a strong magnetic interaction between the hard and soft phases is achieved when the mean grain size of the phases is about 10 nm, corresponding to the ferromagnetic correlation length. Recently, Chen *et al.* [4] and Ahmed *et al.* [5] reported that the addition of Nb prevents the increase of the grain size of the $\alpha$-Fe phase produced by heat treatment. In this paper, we report preliminary results of $Nd_2(Fe_{100-x}Nb_x)_{14}B$ prepared by mechanical alloying. The magnetic and structural properties of milled powders with $x = 0$ and $x = 4$ have been investigated over a wide range of milling times.

## 2. Experimental procedure

Samples of the $Nd_2(Fe_{100-x}Nb_x)_{14}B$ system with $x = 0$ and $x = 4$ were prepared by mechanical alloying using high-purity neodymium (−40 mesh), iron (58 nm), boron (2 μm) and niobium (−325 mesh) powders with appropriate contents, in a Fritsch planetary ball mill of high energy at 280 rpm. The milling was made in hardened stainless steel vials provided with five balls of the same material with 1.5-cm diameter. The ball to powder mass ratio used was 20 : 1. The alloying times were 6, 12, 24, 48, 96, 144, 192 and 240 h. Room-temperature Mössbauer spectra were recorded in a conventional constant acceleration transmission spectrometer with a $^{57}$Co-/Rh source. The Mössbauer spectra of these samples were fitted using the Mosfit program [6] and $\alpha$-Fe as calibration sample. The X-ray patterns were performed at room temperature for all samples using a diffractometer with the Cu K$\alpha$ radiation and the diffractograms were fitted with the Maud program, procedure which is based on the Rietveld method combined with Fourier transform analysis [7]. The magnetic properties of the powders were measured with a vibrating sample magnetometer (VSM) in fields up to 70 kOe using a superconducting quantum interference device (SQUID) magnetometer.

## 3. Results and discussion

The Mössbauer spectra obtained at room temperature (RT) on the $Nd_2(Fe_{100-x}Nb_x)_{14}B$ powders ($x = 0$ and 4) milled for 6, 48, 96, 144 and 240 h, are shown in Figure 1. One observes magnetic sextets composed of lines, the broadening of which increases when the milling time increases. It is important to emphasize that the profile of outermost lines is asymmetrical but the outer wing remains sharp. Consequently, those spectra were first fitted by means of a distribution of hyperfine fields and then by means of two components: (i) a single sextet attributed to an iron phase ($\alpha$-Fe) and (ii) a hyperfine field distribution (HFD) assigned to a disordered iron-based phase.

It can be observed that the samples with and without Nb, whatever the milling time is, show similar hyperfine structures. The disordered iron-based

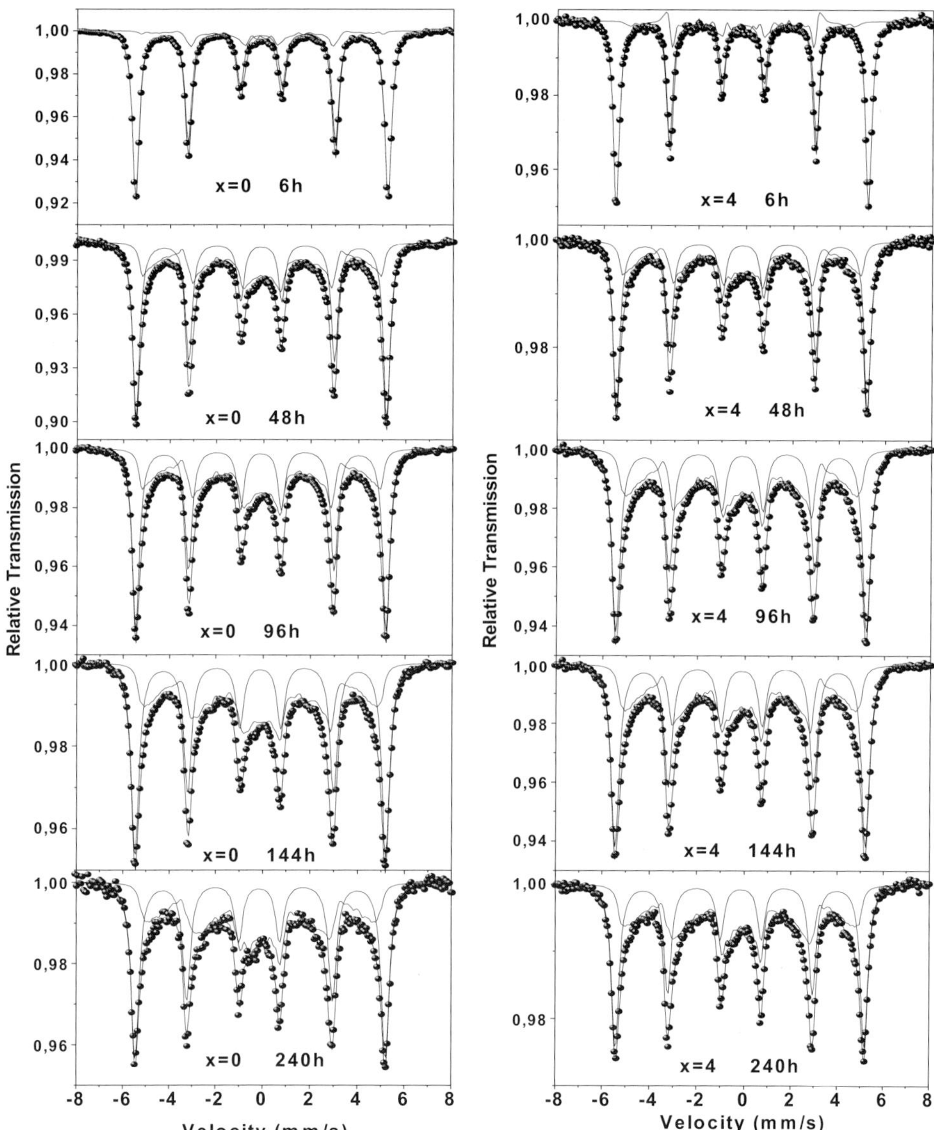


*Figure 1.* Mössbauer spectra of $Nd_2(Fe_{100-x}Nb_x)_{14}B$ samples with $x = 0$ and $x = 4$ for milling times of 6, 48, 96, 144 and 240 h.

phase increases to the expense of the ($\alpha$-Fe) phase, when the alloying time increases, as is shown in Figure 2. In addition, it is important to emphasize that for milling times of about 240 h, the Fe content of the two phases is rather similar.

Figure 3 compares both the hyperfine field (HF) and the mean hyperfine field (MHF) as a function of milling time, for the $\alpha$-Fe and the disordered Fe-based

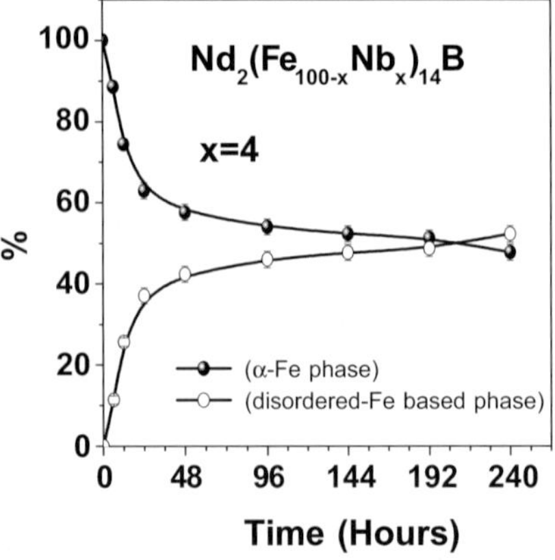


*Figure 2.* Variation of the spectral area of the phases during mechanical milling.

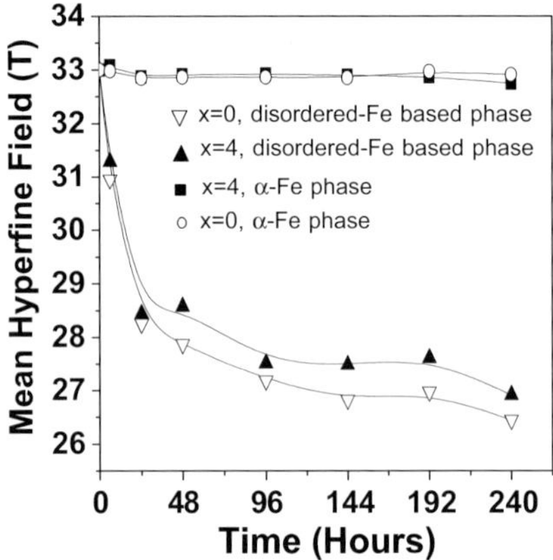


*Figure 3.* Dependence of the Mean Hyperfine Field with the milling time.

phases for $x = 0$ and $x = 4$, respectively. It can be observed that the HF of the $\alpha$-Fe remains nearly constant for the two Nb contents, whereas for the disordered Fe-based phase the MHF decreases with the milling time up to 48 h and then it nearly stabilizes. Also it can be noted that a slightly higher MHF is obtained for samples with $x = 4$, due to the presence of Nb.

The X-ray diffractograms of the samples are shown in Figure 4. It can be seen, in all the patterns, the peaks of $\alpha$-Fe and some additional peaks located at $2\theta$ values of 30.3°, 35.3°, 50.4° and 60.0° which can be attributed to a disordered iron-based phase. It can be also noted the progressive broadening of the peaks as the milling time increases. Then these results show that the HFD obtained by Mössbauer spectrometry can be assigned to this disordered iron-based phase.

The right pattern of Figure 4, corresponding to sample with $x = 4$ and 6 h milling, presents additional peaks at $2\theta = 38.5°$ and 55.8° which correspond to

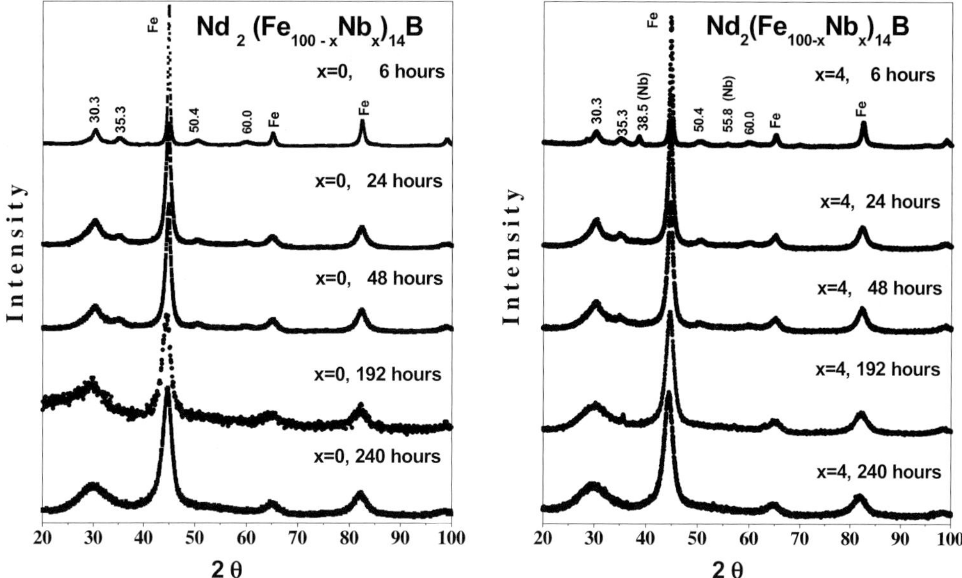


*Figure 4.* X-ray diffractograms for the studied samples.

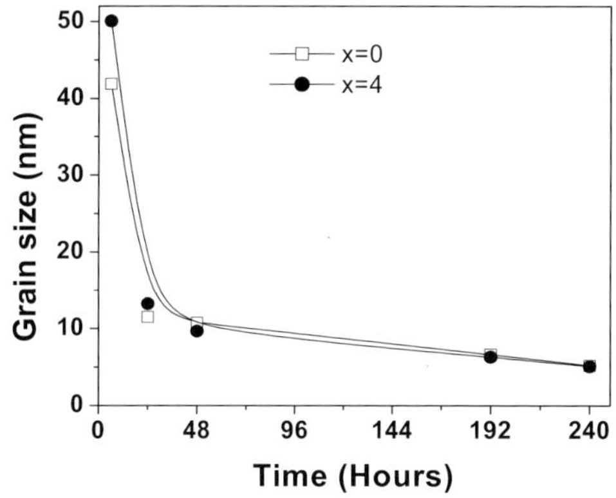


*Figure 5.* Dependence of the mean grain size with the milling time.

those characteristic of Nb. These peaks disappear for larger milling times, suggesting that the alloying is effective. The lattice parameter obtained for the $\alpha$-Fe phase remains nearly constant ($\approx$2.87 Å ). This behavior in conjunction with the constant value of the HF suggest to conclude that the $\alpha$-Fe phase remains nearly pure whatever the alloying time is increasing. Then, the Nd, B and Nb atoms must be dispersed into the disordered iron-based phase.

Figure 5 shows the dependence of the mean grain size of the $\alpha$-Fe phase with the milling time for samples with and without Nb. It can be observed a rapid

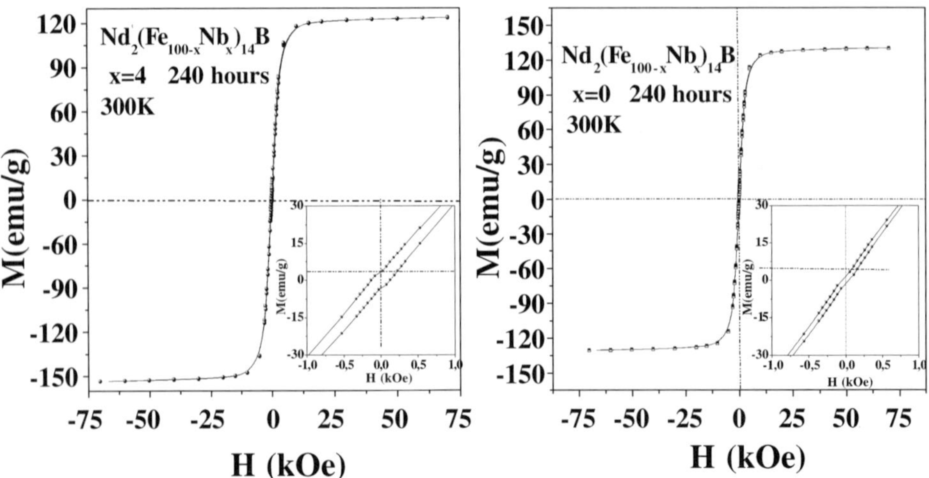


*Figure 6.* Typical hysteresis loop of the $Nd_2(Fe_{100-x}Nb_x)_{14}B$ mechanically alloyed powders for $x = 4$ and $x = 0$.

decrease of the grain size up to 48 h and then a slow decrease with the alloying time. Also, it can be noted that the Nb presence practically does not affect the $\alpha$-Fe grain size.

A typical hysteresis loop of one of the milled samples is shown in Figure 6. It can be observed a rather low coercive field, providing thus to all these nanostructured powders a soft ferromagnetic behaviour. Such a result confirms that no hard magnetic phase is created during the alloying process, in agreement with both X-ray diffraction and Mössbauer analysis. The results for $x = 4$ and $x = 0$ are very similar, like one observed in Figure 6.

In conclusion, mechanical alloying has been used to prepare $Nd_2(Fe_{100-x}Nb_x)_{14}B$ nanostructured powders, with $x = 0$ and $x = 4$. They consist of bcc-Fe nanograins dispersed into a highly disordered iron-based phase, and exhibiting a magnetically soft behavior. High-temperature annealings under different conditions are now in progress in order to obtain the magnetically hard phase, and we hope that the Nb addition avoids the growth of soft phase as it is reported in [4, 5].

## Acknowledgements

We are grateful to the Central Committee of Research of the University of Tolima for its financial support, to the Mössbauer Laboratory of the University of Valle, and the Ecos-Nord Program (C03P01), for the achievement of this work. We are grateful to A. M. Mercier from Laboratoire des Fluorures of Université du Maine UMR CNRS 6010 for performing XRD measurements and to the Laboratory of Magnetism of Bilbao-Spain for the VSM measurements.

## References

1. Croat J. J., Herbst J. F., Lee R. W. and Pinkerton F. E., *J. Appl. Phys.* **55** (1984), 2078.
2. Hadjipanayis G. C., Hazelton R. C. and Lawless K. R., *J. Appl. Phys.* **55** (1984), 2073.
3. Kneller E. F. and Hawig R., *IEEE Trans. Magn.* **27** (1991), 3588.
4. Chen Z., Zhang Y., Ding Y., Hadjipanayis G. C., Chen Q. and Ma B., *J. Magn. Magn. Mater.* **195** (1999), 420.
5. Ahmed F. M., Ataie A., Williams A. J. and Harris I. R., *J. Magn. Magn. Mater.* **157** (1996), 59.
6. Bremers H., Jarms C., Hesse J., Chajwasiliou S., Efthiadis K. G. and Tsonkalas Y., *J. Magn. Magn. Mater.* **140** (1995), 63.
7. Lutterotti L. and Scardi P., *J. Appl. Crystallogr.* **23** (1990), 246.

Hyperfine Interactions (2005) 161:211–220
DOI 10.1007/s10751-005-9193-1


# Mechanosynthesis of YIG and GdIG: A Structural and Mössbauer Study

A. PAESANO JR.[1,*], S. C. ZANATTA[1], S. N. DE MEDEIROS[1], L. F. CÓTICA[1] and J. B. M. DA CUNHA[2]
[1]*Departamento de Física, Universidade Estadual de Maringá, Av. Colombo 5790, CEP 87020-900, Maringá, PR, Brazil; e-mail: paesano@wnet.com.br*
[2]*Instituto de Física, Universidade Federal do Rio Grande do Sul, Porto Alegre, Brazil*

**Abstract.** We have investigated the mechanosynthesis of gadolinium and yttrium iron garnets by high-energy ball-milling of $\alpha$-$Fe_2O_3$ and $Gd_2O_3$ or $\alpha$-$Fe_2O_3$ and $Y_2O_3$, respectively, followed by short thermal annealings conducted at moderate temperatures. The samples were characterized by X-ray diffraction and Mössbauer spectroscopy, in order to determine the influence of the milling time and annealing conditions on the final products. For as-milled samples of each rare-earth system, the results revealed the formation of perovskite phases, in relative amounts that depend on the milling time. The formation of garnet phases was observed in as-annealed samples treated at 1000°C for 2 h or 1100°C for 3 h, i.e., at very modest annealing requirements when compared with ordinary solid-state-reaction processes performed without previous high-energy milling. Also, the occurrence was verified of a milling time for which the relative amount of garnet phases formed by annealing was maximized. This time depends on the rare-earth composing the garnet phase and on the annealing temperature.

**Key Words:** gadolinium-iron garnet (GdIG), high-energy ball-milling, Mössbauer spectroscopy, yttrium-iron garnet (YIG).

## 1. Introduction

Recent exponential growth in microwave communication through mobile and satellite communications has stressed the worldwide need for extremely low loss and economical microwave devices using ferrite materials. The performance of these devices was meaningfully improved with the discovery of the yttrium iron garnet (YIG), in 1956 [1]. To date, YIG is considered the best microwave material in the 1–10 GHz band [2]. Differentiated magnetic properties can also be obtained by substituting yttrium or iron for other elements. The gadolinium iron garnet (GdIG), for instance, can offer superior temperature

* Author for correspondence.

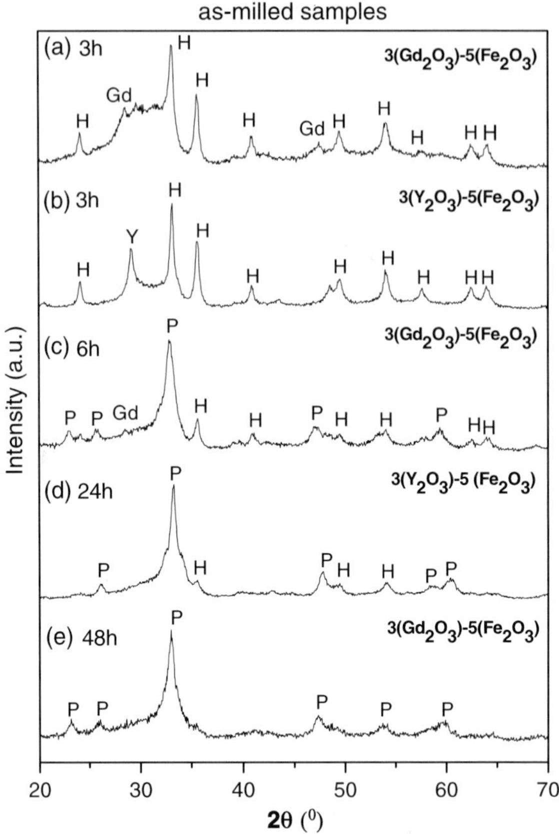


*Figure 1.* X-ray diffraction patterns for $3(RE_2O_3)$–$5(Fe_2O_3)$ as-milled samples; *H* = $Fe_2O_3$; *Gd* = $Gd_2O_3$; *Y* = $Y_2O_3$; *P* = $GdFeO_3/YFeO_3$.

stability and increases the power handling in comparison with the yttrium iron garnet [2].

Both YIG and GdIG have been extensively investigated because of their technological importance, particularly in the conception of microwave equipment [3, 4]. In most cases, these materials, frequently used by industry, are sintered polycrystalline ceramics, prepared by solid-state-reaction of the parent oxides [5]. The process usually requires a calcination temperature of over 1400°C and this high temperature inevitably leads to coarsened microstructures, which affect the sinterability of the synthesized powders.

Aiming to overcome this constraint, some alternative routes for garnet preparation such as coprecipitation, sol–gel, hydrothermal synthesis, spray and freeze-drying have been investigated. Among others, high-energy ball-milling or mechanochemical processing has been successfully used to synthesize a wide range of nanosized ceramic powders. In addition, the ball-milling process can also enhance the reaction of multi-component systems by significantly lowering the calcination temperatures.

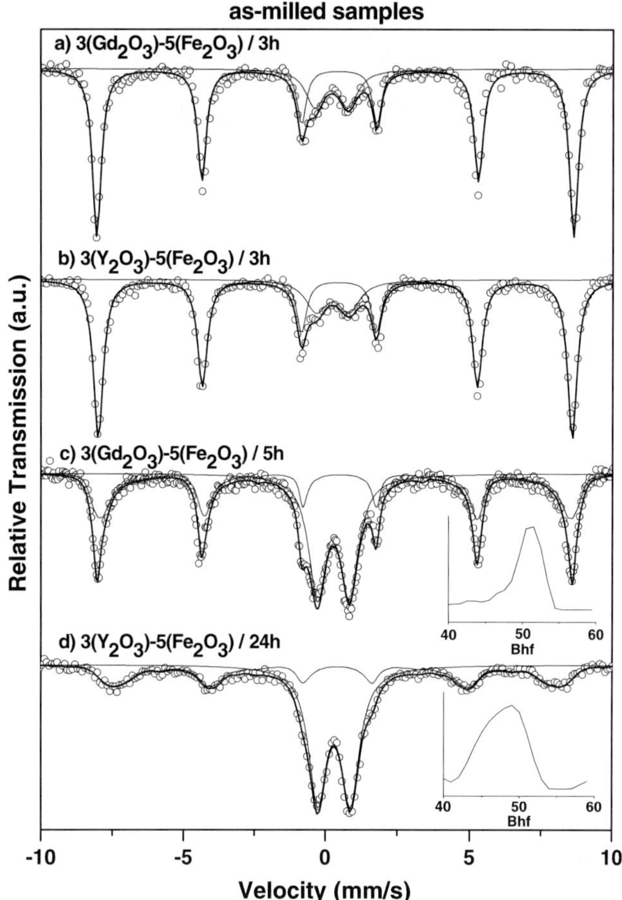


*Figure 2.* Room-temperature Mössbauer spectra of as-milled samples: a) $3(Gd_2O_3)$–$5(Fe_2O_3)$–3 h, b) $3(Y_2O_3)$–$5(Fe_2O_3)$–3 h, c) $3(Gd_2O_3)$–$5(Fe_2O_3)$–5 h and d) $3(Y_2O_3)$–$5(Fe_2O_3)$–24 h; (*insets*: magnetic hyperfine field distributions).

In this sense, we have investigated the formation of garnets by employing high-energy ball-milling as a prior step to heat treatments. In this paper, we report X-ray diffraction (XRD) and Mössbauer spectroscopy (MS) results obtained on the synthesis of $Gd_3Fe_5O_{12}$ and $Y_3Fe_5O_{12}$ garnet phases obtained in moderate annealing conditions.

## 2. Experimental details

$Gd_3Fe_5O_{12}$ and $Y_3Fe_5O_{12}$ compounds were prepared, at the start, by high-energy ball-milling analytical grade oxides, $Fe_2O_3$ and $Y_2O_3$ or $Fe_2O_3$ and $Gd_2O_3$, respectively, manually pre-mixed in nominal compositions of 5:3. The precursors were dry-milled for times ranging between 1 h and 48 h, in argon atmosphere, in

*Table I.* Mössbauer hyperfine parameters for $3(RE_2O_3)–5(Fe_2O_3)$ as-milled samples

| Milling time (h) | | Subspectrum | $B_{Hf}$[a](T) | $\Delta E_Q$ (mm/s) | IS[b] (mm/s) | Area (%) |
|---|---|---|---|---|---|---|
| $3(Gd_2O_3)–5(Fe_2O_3)$ | | | | | | |
| 3 | | Doublet | – | 1.13 | 0.24 | 20.0 |
| | | Sextet | 51.4 | −0.18 | 0.37 | 80.0 |
| 5 | | Doublet | – | 1.16 | 0.18 | 43.8 |
| | | Sextet | 50.3 | −0.17 | 0.38 | 56.2 |
| 12 h | RT | Doublet | – | 1.13 | 0.29 | 55.0 |
| | | Dist. | 49.8 | −0.17 | 0.36 | 45.0 |
| | 85 K | $Fe_2O_3$ | 52.9 | −0.24 | 0.44 | 18.0 |
| | | $GdFeO_3$ | 53.1 | 0.02 | 0.50 | 38.9 |
| | | Doublet | – | 1.27 | 0.38 | 43.1 |
| | 5 K | $Fe_2O_3$ | 54.0 | −0.28 | 0.44 | 18.0 |
| | | $GdFeO_3$ | 54.3 | 0.11 | 0.51 | 37.6 |
| | | Dist. | 46.7 | −0.02 | 0.33 | 44.4 |
| 24 | | Doublet | – | 1.10 | 0.31 | 64.0 |
| | | Dist. | 48.6 | −0.10 | 0.38 | 36.0 |
| 48 | | Doublet | – | 1.10 | 0.30 | 75.0 |
| | | Dist. | 48.0 | −0.11 | 0.36 | 25.0 |
| $3(Y_2O_3)–5(Fe_2O_3)$ | | | | | | |
| 3 | | Doublet | – | 1.16 | 0.23 | 17.9 |
| | | Sextet | 51.4 | −0.17 | 0.37 | 82.1 |
| 12 | | Doublet | – | 1.11 | 0.29 | 63.2 |
| | | Dist. | 50.9 | −0.23 | 0.37 | 48.8 |
| 24 | | Doublet | – | 1.14 | 0.30 | 64.0 |
| | | Dist. | 44.5 | −0.10 | 0.37 | 36.0 |

[a] Average hyperfine magnetic field, in case of distribution.
[b] Relative to $\alpha$-Fe foil at room temperature.
Obs. The linewidth used in the magnetic distributions and discrete sextets varied between 0.27 and 0.30 mm/s.

a Fritsch Pulverisette 6 planetary ball-mill, using a 80 $cm^3$ charged with 10 mm diamenter stainless-steel balls, and under closed milling conditions, i.e., the vial was not opened during the milling process. The ball-to-powder mass ratio (20:1) and angular velocity of the supporting disc and vial (31.42 rad $s^{-1}$) were kept constant throughout the experiments. Subsequent to the milling, the samples were annealed in free atmosphere at 1000°C and 1100°C for 2 h and 3 h, respectively.

The XRD patterns of the as-milled and annealed products were obtained at room temperature (RT), using a Siemens D500 X-ray diffractometer in Bragg–Brentano geometry, with Cu $K_\alpha$ radiation, in the $20° \leq 2\theta \leq 70°$ range. The MS characterization was performed in the transmission geometry, using a conventional Mössbauer spectrometer in a constant acceleration mode. The $\gamma$-rays were provided by a nominal 15 mCi $^{57}Co(Rh)$ source. The Mössbauer spectra were analyzed with a non-linear least-square routine, with Lorentzian line shapes. The values of isomer shift (IS) are referred to that of $\alpha$-Fe foil at RT.

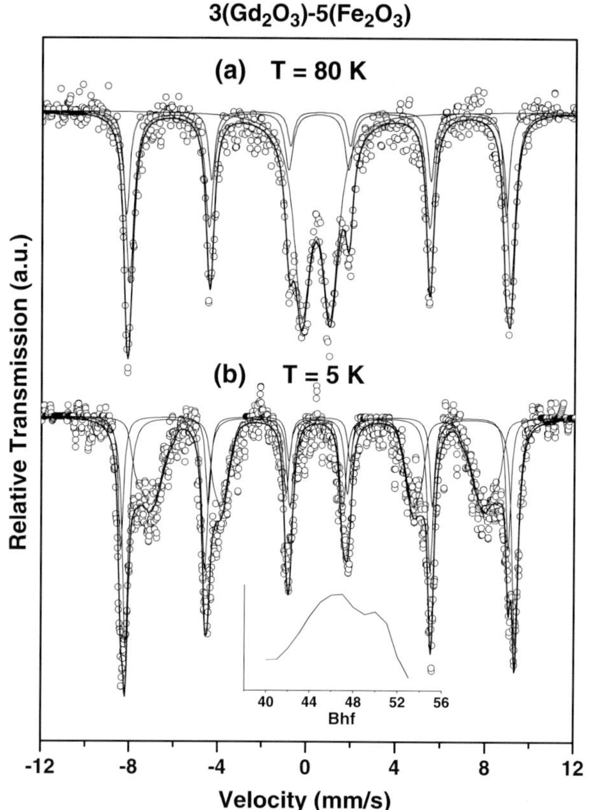


*Figure 3.* Mössbauer spectra obtained at low temperature for the $3(Gd_2O_3)–5(Fe_2O_3)$ 12 h as-milled sample: (a) $T = 80$ K, (b) $T = 5$ K; (*inset*: magnetic hyperfine field distribution).

## 3. Results and discussion

Figure 1 shows the XRD diffraction patterns for some selected $3(Gd_2O_3)–5(Fe_2O_3)$ and $3(Y_2O_3)–5(Fe_2O_3)$ as-milled samples.

The diffractograms show no clear phase formation until 3 h of milling (Figure 1a and b), although the reflection peaks for $Fe_2O_3$, $Y_2O_3$ and $Gd_2O_3$ are largely broadened and their intensities are strongly reduced, as a consequence of the grain size refining and very disordered or defected structures. The XRD profile of the 6 h milled gadolinium containing sample (Figure 1c) still shows the most intense peaks of hematite and gadolinia, and new peaks corresponding to reflection planes of the $GdFeO_3$ perovskite compound. After 24 h, or more, of milling only perovskites ($YFeO_3/GdFeO_3$), as majority phases, and traces of precursors could be identified in the diffractograms of both rare-earth systems (Figure 1d and e). It was also noted that no evidence for the presence of a garnet can be observed.

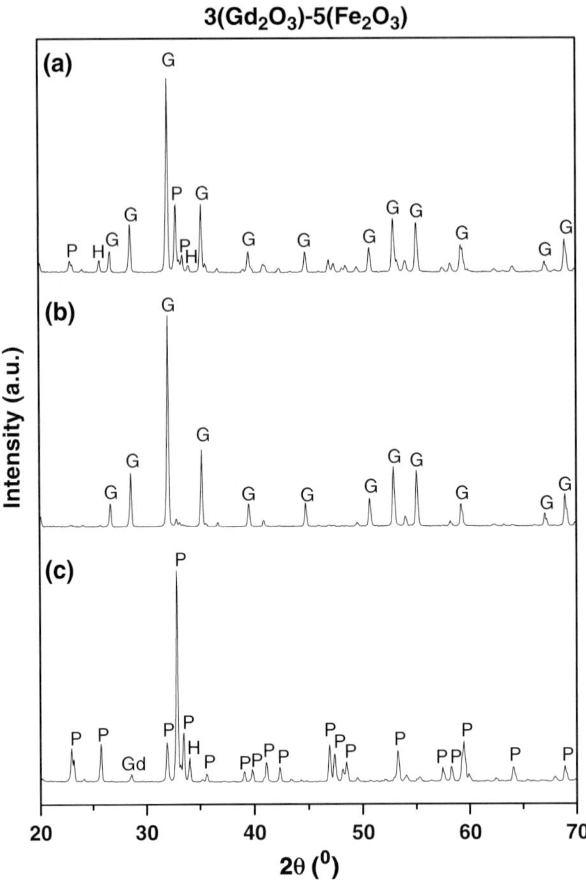


*Figure 4.* X-ray diffraction patterns for $3(Gd_2O_3)$–$5(Fe_2O_3)$ milled or annealed samples: (a) 6 h milled (1000°C/2 h), (b) 12 h milled (1100°C/3 h) and (c) un-milled (1100°C/3 h); $G = Gd_3Fe_5O_{12}$; $P = GdFeO_3$.

The Mössbauer spectra obtained at RT for both as-milled rare-earth systems are presented in Figure 2.

After 3 h of milling, the hematite magnetic pattern and a paramagnetic component were revealed. The fitted Mössbauer parameters, for these and other samples, are listed in Table I.

For longer milling times (i.e., $t \geq 5$ h), Mössbauer spectra exhibited a hyperfine field distribution in addition to the earlier doublet, which increased with the milling time. At this point, the unambiguous assignment of these two contributions is far from straight since it could belong to magnetic (distribution) or superparamagnetic (doublet) nanoparticles of $\alpha$-$Fe_2O_3$, $GdFeO_3$ or, even, $Gd_3Fe_5O_{12}$.

Aiming to better resolve these spectra, a few samples, among the longer as-milled ones, were measured at low temperatures (i.e., $T \leq 80$ K). Figure 3 shows

*Table II.* Mössbauer hyperfine parameters for $3(RE_2O_3)$–$5(Fe_2O_3)$ milled and annealed samples; (a) 1000°C/2 h–(b) 1100°C/3 h

| Milling time (h) | | Subspectrum/site | $B_{Hf}$[a] (T) | $\Delta E_Q$ (mm/s) | IS[b] (mm/s) | Area (%) |
|---|---|---|---|---|---|---|
| $3(Gd_2O_3)$–$5(Fe_2O_3)$ | | | | | | |
| 1 | (a) | Garnet/16(a) | 49.7 | 0.06 | 0.38 | 32.8 |
| | | Garnet/24(d) | 40.4 | 0.00 | 0.16 | 49.2 |
| | | $Fe_2O_3/GdFeO_3$ | 51.5 | −0.14 | 0.32 | 18.0 |
| | (b) | Garnet/16(a) | 49.4 | 0.06 | 0.38 | 40.0 |
| | | Garnet/24(d) | 40.3 | 0.00 | 0.16 | 60.0 |
| 3 | (a) | Garnet/16(a) | 49.9 | 0.04 | 0.38 | 36.5 |
| | | Garnet/24(d) | 40.8 | 0.00 | 0.16 | 54.7 |
| | | $Fe_2O_3/GdFeO_3$ | 51.8 | −0.11 | 0.32 | 8.8 |
| | (b) | Garnet/16(a) | 49.3 | 0.04 | 0.38 | 40.0 |
| | | Garnet/24(d) | 40.3 | 0.01 | 0.17 | 60.0 |
| 5 | (a) | Garnet/16(a) | 49.9 | 0.14 | 0.30 | 29.8 |
| | | Garnet/24(d) | 40.4 | 0.01 | 0.16 | 44.8 |
| | | $Fe_2O_3/GdFeO_3$ | 50.6 | −0.19 | 0.32 | 25.4 |
| | (b) | Garnet/16(a) | 49.3 | 0.04 | 0.38 | 40.0 |
| | | Garnet/24(d) | 40.3 | 0.01 | 0.17 | 60.0 |
| 6 | (a) | Garnet/16(a) | 50.2 | 0.13 | 0.40 | 20.4 |
| | | Garnet/24(d) | 40.5 | 0.00 | 0.16 | 30.7 |
| | | $Fe_2O_3/GdFeO_3$ | 51.0 | −0.12 | 0.35 | 48.9 |
| | (b) | Garnet/16(a) | 49.5 | 0.03 | 0.39 | 40.0 |
| | | Garnet/24(d) | 39.6 | 0.03 | 0.16 | 60.0 |
| 12 | (a) | $Fe_2O_3$ | 51.0 | −0.30 | 0.37 | 43.1 |
| | | $GdFeO_3$ | 50.5 | 0.12 | 0.35 | 56.9 |
| | (b) | Garnet/16(a) | 49.3 | 0.02 | 0.37 | 40.0 |
| | | Garnet/24(d) | 40.3 | 0.01 | 0.16 | 60.0 |
| 24 | (a) | $Fe_2O_3$ | 51.1 | −0.37 | 0.34 | 45.0 |
| | | $GdFeO_3$ | 50.6 | 0.12 | 0.38 | 55.0 |
| | (b) | Garnet/16(a) | 49.6 | 0.00 | 0.38 | 35.3 |
| | | Garnet/24(d) | 40.3 | 0.00 | 0.18 | 53.0 |
| | | $Fe_2O_3/GdFeO_3$ | 51.9 | −0.17 | 0.34 | 11.7 |
| $3(Y_2O_3)$–$5(Fe_2O_3)$ | | | | | | |
| 3 | (a) | $Fe_2O_3$ | 51.3 | −0.28 | 0.38 | 36.7 |
| | | $YFeO_3$ | 49.7 | 0.13 | 0.34 | 63.3 |
| 6 | (a) | $Fe_2O_3$ | 51.1 | −0.34 | 0.32 | 39.4 |
| | | $YFeO_3$ | 50.4 | 0.09 | 0.38 | 60.6 |
| | (b) | Garnet/16(a) | 49.5 | 0.02 | 0.39 | 40.0 |
| | | Garnet/24(d) | 39.6 | 0.06 | 0.16 | 60.0 |
| 12 | (a) | $Fe_2O_3$ | 50.9 | −0.36 | 0.31 | 43.4 |
| | | $YFeO_3$ | 50.5 | 0.13 | 0.37 | 56.6 |
| | (b) | Garnet/16(a) | 48.9 | 0.00 | 0.39 | 38.0 |
| | | Garnet/24(d) | 39.3 | 0.08 | 0.16 | 55.5 |
| | | $Fe_2O_3/YFeO_3$ | 49.8 | −0.02 | 0.38 | 6.5 |
| 24 | (a) | $Fe_2O_3$ | 51.1 | −0.37 | 0.34 | 45.0 |
| | | $YFeO_3$ | 50.6 | 0.12 | 0.38 | 55.0 |
| | (b) | Garnet/16(a) | 48.9 | 0.04 | 0.38 | 35.5 |
| | | Garnet/24(d) | 39.4 | 0.09 | 0.14 | 53.2 |
| | | $Fe_2O_3/YFeO_3$ | 51.7 | −0.06 | 0.39 | 11.3 |

[a] Average hyperfine magnetic field, in case of distribution.
[b] Relative to $\alpha$-Fe foil at room temperature.
Obs. The linewidth used in the magnetic distributions and discrete sextets varied between 0.27 and 0.30 mm/s.

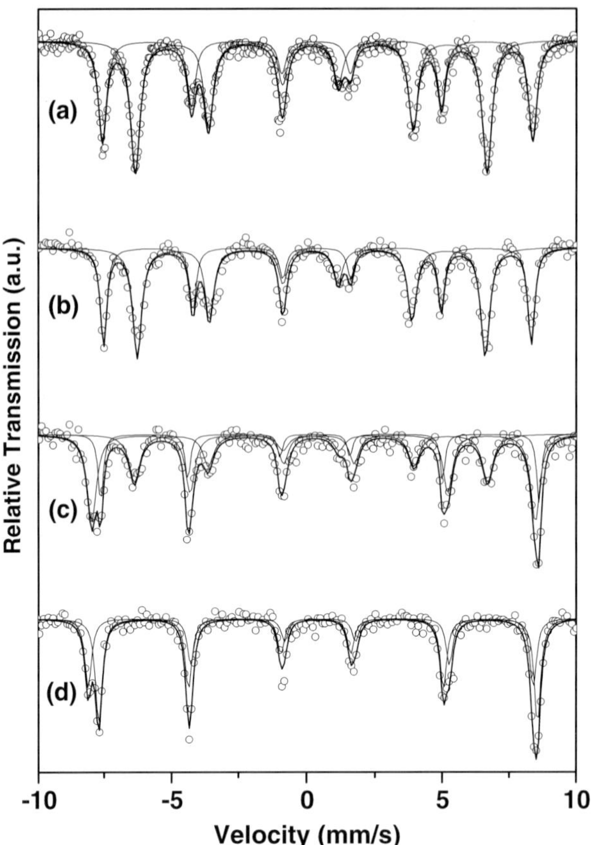


*Figure 5.* Room-temperature Mössbauer spectra for milled and annealed samples: (a) $3(Gd_2O_3)$–$5(Fe_2O_3)$ 5 h milled, annealed at 1100°C for 3 h; (b) $3(Y_2O_3)$–$5(Fe_2O_3)$ 6 h milled, annealed at 1100°C for 3 h; (c) $3(Gd_2O_3)$–$5(Fe_2O_3)$ 6 h milled, annealed at 1000°C for 2 h and (d) $3(Y_2O_3)$–$5(Fe_2O_3)$ 6 h milled, annealed at 1000°C for 2 h.

the spectra obtained specifically for the 12 h-milled gadolinium containing sample, measured at 80 K (Figure 3a) and 5 K (Figure 3b).

The spectra obtained in this temperature range could be fitted using three components: two discrete sextets belonging to hematite and perovskite phases plus either a doublet, in the case of 80 K spectrum, or a hyperfine magnetic field distribution, in the case of the 5 K spectrum. No constraints were imposed to the fits. In decreasing the temperature up to 5 K, the doublet was totally converted in a magnetic field distribution due to a progressive lowering of the relaxation frequency of some assembly of superparamagnetic particles.

The X-ray diffractograms of some $3(Gd_2O_3)$–$5(Fe_2O_3)$ annealed samples are shown in Figure 4.

The patterns for the 1000°C/2 h and 1100°C/3 h annealed samples, previously milled for 6 h and 12 h, respectively, (Figure 4a and b) revealed peaks unequivocally identified in the respective diffraction spectra as being for

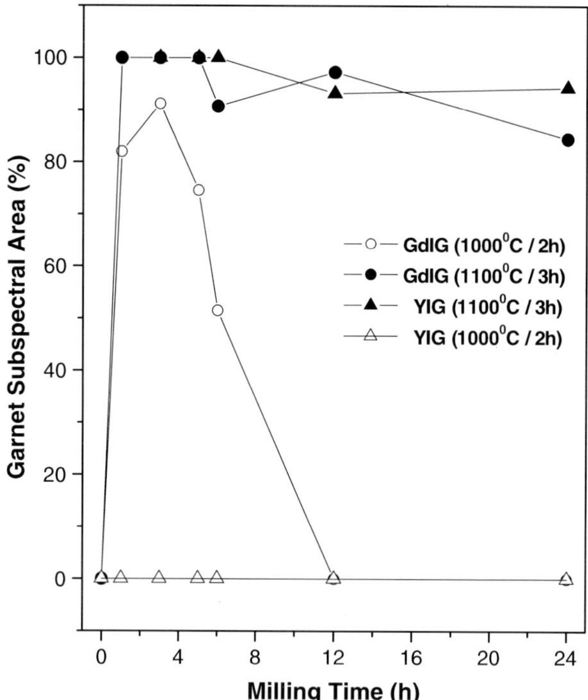


*Figure 6.* Mössbauer relative areas for $Gd_3Fe_5O_{12}$ (*circles*) and $Y_3Fe_5O_{12}$ (*triangles*) phases present in annealed samples, as a function of milling time.

GdIG. For comparison, the diffractogram of an unmilled gadolinium containing sample annealed at 1100°C for 3 h is shown in Figure 4c, where no garnet phase could be distinguished. Diversely, for yttrium samples the garnet formation was observed only for the 1100°C/3 h thermal treatment (spectrum not shown), revealing a lower crystallization ability than the one for the gadolinium system.

The Mössbauer spectra of some annealed samples are presented in Figure 5 and the respective hyperfine parameters displayed in Table II.

It can be verified, either in the gadolinium compound spectrum (Figure 5a) or in the yttrium compound spectrum (Figure 5b), for the 1100°C/3 h annealed samples, the presence of particular pairs of sextets, having magnetic hyperfine fields comparable to those values previously reported in literature for GdIG or YIG [6, 7]. The subspectral areas were constrained in the fitting routine according to the respective populations of the two iron sites in garnets (i.e., 2:3) [8].

Hence, consistent with the X-ray results, these Mössbauer spectra revealed in each case the formation of, basically, a garnet single phase. For the gadolinium compound milled by 6 h and then annealed at 1000°C for 2 h, besides the garnet spectral components, an unresolved magnetic contribution could be observed belonging, most probably, to hematite or perovskite. The latter are the only (resolved) components present in the spectrum of the 1000°C/2 h annealed yttrium compound (Figure 5d), thus verifying, once again in agreement with

XRD results (not shown), that no yttrium garnet formation took place in this annealing condition.

Table II reveals that the hyperfine parameters of the annealed samples do not show any significant variation over the (previous) milling time, although the relative area for each garnet phase does show that, as displayed in Figure 6.

Interestingly, an unexpected behavior is displayed in the figure and constitutes strong evidence, more clearly for the 1000°C/2 h annealed gadolinium samples, for the existence of an optimal milling time after which the amount of garnet phase thermally formed is maximized.

## 4. Conclusions

The GdIG and YIG phases were successfully prepared by annealing high-energy ball-milled parent oxides, in times and temperatures significantly lower when compared with solid-state-reaction processes carried out directly. The notable enhancement in the formation of these garnet phases can be attributed to an effective particle size reduction, due to the high-energy ball-milling as a step prior to annealing. In addition, immediately after milling, it was observed the mecanosynthesis of perovskite, which is usually presented as a transient phase before the garnet phase is finally formed. The occurrence was also verified of a milling time for which the relative amount of garnet phases formed by annealing was maximized. This time depends on the rare-earth composing the garnet phase and on the annealing temperature. For a large scale garnet production, this certainly constitutes key information.

## Acknowledgements

We greatly appreciate the financial support from the Brazilian agencies CAPES (PROCAD), Fundação Araucária and FAPERGS.

## References

1. Sugimoto M., *J. Am. Ceram. Soc.* **82**(2) (1999), 269.
2. Pardavi-Horvath M., *J. Magn. Magn. Mater.* **215–216** (2000), 171.
3. Tsay C.-Y., Liu C.-Y., Liu K.-S., Lin N., Hu L.-J. and Yeh T.-S., *Mater. Chem. Phys.* **79** (2003), 138.
4. Ristic M., Czakó-Nagy I., Popovic S., Music S. and Vérts A., *J. Mol. Struct.* **410–411** (1997), 281.
5. Grosseau P., Bachiorrini A. and Guilhot B., *Powder Technol.* **93** (1997), 247.
6. Ristic M., Felner I., Nowik I., Popovic S., Czakónagy I. and Music S., *J. Alloys Compd.* **308** (2000), 301.
7. Ristic M., Nowik I., Popovic S., Felner I. and Music S., *Mater. Lett.* **57** (2003), 2584.
8. Lataifeh M. S. and Lehlooh A.-F. D., *Solid State Commun.* **97** (1996), 805–807.

Hyperfine Interactions (2005) 161:221–227
DOI 10.1007/s10751-005-9194-0 

# Mössbauer Spectroscopy, Dilatometry and Neutron Diffraction Detection of the $\varepsilon$-Phase Fraction in Fe–Mn Shape Memory Alloys

J. MARTÍNEZ[1], G. AURELIO[2], G. CUELLO[3], S. M. COTES[1], A. FERNÁNDEZ GUILLERMET[2] and J. DESIMONI[1,*]

[1]*Departamento de Física, Facultad de Ciencias Exactas, Universidad Nacional de La Plata, IFLP-CONICET, C.C.6 7, 1900 La Plata, Argentina; e-mail: desimoni@fisica.unlp.edu.ar*

[2]*Instituto Balseiro, Centro Atómico Bariloche, CNEA and CONICET, Avda. Bustillo 10000, 8400 S. C. de Bariloche, Argentina*

[3]*Diffraction Group, Institut Laue-Langevin, 6 Rue Jules Horowitz, B.P. 15638042 Grenoble Cedex 9, France*

**Abstract.** The results of a Mössbauer spectroscopy (MS) and a neutron diffraction (ND) study carried out on a set of Fe–Mn alloys quenched from high temperatures are reported. Upon quenching the high temperature stable phase FCC ($\gamma$) together with metastably retained HCP-($\varepsilon$) phase are formed. The lattice parameters obtained using ND are in excellent agreement with previous results of the literature. The phase fractions obtained from MS are coincident, within experimental errors, with those extracted from ND, with exception of the case of an alloy with the lowest Mn content (16.8 at.% Mn). That situation can be associated with the occurrence of a high density of stacking faults (SF) in the $\gamma$ phase or to the presence of a paramagnetic $\gamma$ phase. $\varepsilon$ phase fractions obtained from MS and ND decrease smoothly with Mn content but are always larger than the values reported in literature resulting from dilatometry.

## 1. Introduction

The technological applications of shape-memory alloys based in the Fe–Mn systems have motivated extensive basic and applied research. In the Fe–Mn system, the shape-memory effect is governed by a reversible martensitic transformation (MT) between the $\gamma$ (FCC) and $\varepsilon$ (HCP) phases. The design of improved shape-memory alloys requires a deep understanding of this MT [1]. In this way, a new thermodynamic description of the relative stabilities between the $\gamma$- and $\varepsilon$-phases in the Fe–Mn system have been reported [2]. This description requires a critical test such as the comparison of the predicted enthalpies of transformation ($\Delta H_{\mathrm{m}}$) with experimental data. In order to correctly evaluate the experimental $\Delta H_{\mathrm{m}}$, the fraction of $\varepsilon$ phase $f^{\varepsilon}$ induced by the MT after quenching

* Author for correspondence.

must be determined independently, since it varies with the content of Mn [3]. Recently, new measurements of $\Delta H_m$ have been obtained [4] by combining differential scanning calorimetry (DSC) with MS and dilatometry measurements. In that work [4], MS yielded a $\varepsilon$-phase fraction about 30% larger than that of dilatometry. This is a serious discrepancy since dilatometry is traditionally used to derive phase fractions from measured volume changes. Anyhow, the model predictions about $\Delta H_m$ better represented the experimental results obtained using the $f^{\varepsilon}$ determined using MS.

The present work intends to clarify the discrepancy between dilatometry and MS results, by applying a third method, namely, a neutron diffraction study of the quenched Fe–Mn alloys in the [10–30 wt.%] range of Mn content. More specifically, we established the phase fractions of $\varepsilon$- and $\gamma$-phases by a full-pattern Rietveld analysis of neutron diffraction spectra taken in the D1B diffractometer at ILL (Grenoble, France), using a rotating sample holder to satisfactorily approach the conditions of a powder diffraction experiment. We present a detailed comparison and discussion of the results obtained by Mössbauer spectroscopy, dilatometry and neutron diffraction.

## 2. Experimental

Alloys in the concentration range from 15 to 30 wt.% Mn were prepared using Fe and Mn, both of 99.98% purity, and melted in an arc furnace, on a water-cooled copper crucible, in a 350 Torr Ar atmosphere. Polycrystalline samples used for ND experiments consisted of about 2–3 g of small cubes of about 2 mm side, cut from the original ingots. Samples for MS had approximately 2.5 cm × 2.5 cm × 0.08 cm size. To obtain the maximum amount of $\varepsilon$ phase, all specimens were annealed for 1 h at 950°C under Ar atmosphere, quenched at room temperature (RT) without breaking the capsule and then cooled down to liquid nitrogen temperature.

Mössbauer spectra were recorded using a 5 mCi $^{57}$CoRh source and a standard 512 channels constant acceleration spectrometer. The samples were analyzed using Conversion Electron Mössbauer Spectroscopy with a constant flux helium–methane detector. Velocity calibration was performed against a 12 μm-thick $\alpha$-Fe foil. All isomer shifts are referred to this standard at 298 K. The spectra were fitted with Lorentzian line shapes with a non-linear least-squares program with constraints. The relative fraction of phases were determined assuming the same Lamb–Mössbauer factor for the different phases.

The neutron diffraction experiments were performed with the D1B two-axis powder diffractometer at the Institut Laue–Langevin, using a Ge monochromator (311 reflection) to obtain a wavelength of 1.28 Å. The flux on the sample was $0.4 \times 10^6$ n cm$^{-2}$s$^{-1}$. This diffractometer is equipped with a $^3$He multidetector containing 400 cells, with an angular span of 80°, providing high quality spectra. The measurements were performed at room temperature using a vanadium

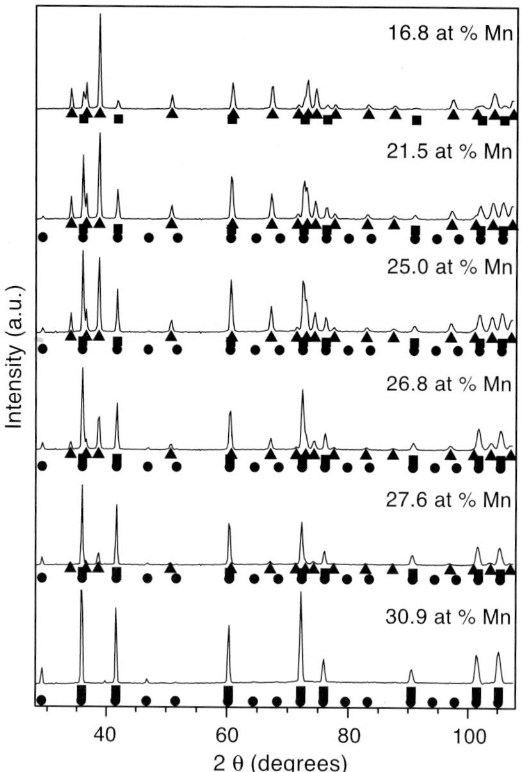


*Figure 1.* Neutron diffractograms taken with D1B instrument at ILL. *Circles*: antiferromagnetic $\gamma$-phase, *squares*: paramagnetic $\gamma$-phase and *triangles*: $\varepsilon$-phase.

cylinder filled with about 2 g of sample and mounted on a rotator device. The rotation of the sample-holder around its axis allows attaining a very good approximation to a powder spectrum [5]. The neutron wavelength was calibrated using an $Al_2O_3$ standard. To establish the fractions of the present phases, a full-pattern Rietveld analysis [6] was performed using the Fullprof 2K software [7].

## 3. Results and discussion

The relative fraction of $\varepsilon$ phase was determined from Mössbauer data analysis $\left(f^{\varepsilon}_{MS}\right)$, and from full-pattern Rietveld analysis of neutron diffraction $\left(f^{\varepsilon}_{ND}\right)$.

Figure 1 presents the recorded experimental diffraction patterns. These diffractograms show reflections associated with the $\gamma$ phase and to the $\varepsilon$ phase, with exception of that corresponding to the alloy with 30.95 wt.% Mn where only the $\gamma$ phase is observed. In several samples, there are also additional lines that can be attributed to the antiferromagnetic ordered $\gamma$ phase. Two fitting procedures were carried out. The first one considered the paramagnetic $\varepsilon$ phase

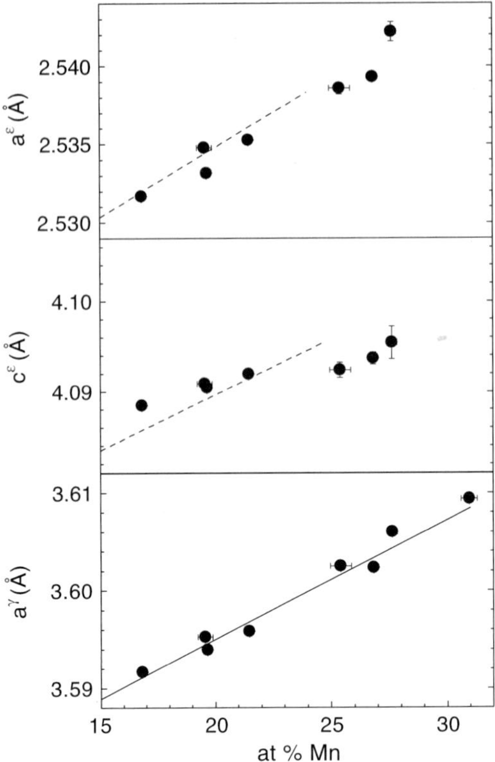


*Figure 2.* Composition dependence of the present $a^{\gamma}$, $a^{\varepsilon}$ and $c^{\varepsilon}$ lattice parameters. *Dashed lines* correspond to ref. [11] and solid line to ref. [12].

contribution in addition to two $\gamma$ phase contributions, one paramagnetic and the other non-collinear antiferromagnetic using the Q3 magnetic model [8–10]. The magnetic moments for Fe and Mn were taken from ref. [10] and were maintained fixed during the fit. The second one was performed taking into account only two contributions to the diffractograms, one corresponding to the paramagnetic $\varepsilon$ phase and the other to the antiferromagnetic $\gamma$ phase using again the Q3 model [8–10], but in this case the parameters related to texture and preferred orientations were permitted to vary during the fit. Both methods satisfactorily reproduced the experimental data within the experimental error and the resulting $f^{\varepsilon}$ differs in less than 3%.

Figure 2 shows the agreement between the lattice parameters for $\gamma$ and $\varepsilon$ phases obtained from the ND results and those taken from literature [11] and [12].

Typical MS spectra for the various alloys are shown in Figure 3. The fact that the ordering Néel temperatures, $T_N^{\gamma}$, are higher than RT [13] and the MS spectra have not a definite structure rules out the possibility to analyze them taking into account the two magnetic states for the $\gamma$ phase, as it could be suggested by the

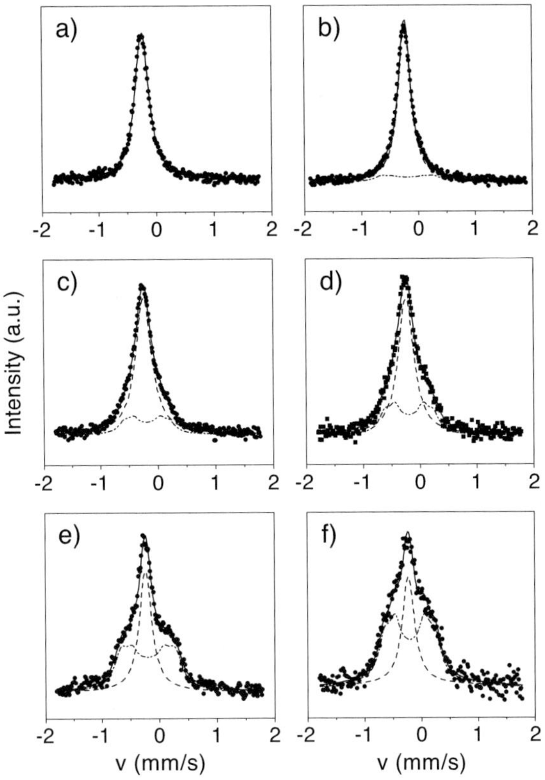


*Figure 3.* Conversion electron Mössbauer spectra recorded on the Fe–Mn alloys. a) 16.8 at.%, b) 19.6 at.%, c) 19.7 at.%, d) 21.5 at.%, e) 25.4 at.%, and f) 27.6 at.%.

*Table I.* Mössbauer hyperfine parameters and relative fractions for the various alloys

| Mn at % | $T_N^\gamma$ (K) | $\delta$(mm/s) | $\Gamma$(mm/s) | $f^\varepsilon$ (%) | $B$(T) | $\delta$(mm/s) | $\Gamma$(mm/s) | $f^\gamma$ (%) |
|---|---|---|---|---|---|---|---|---|
| 16.8 | 318 | $-0.14_1$ | $0.30_1$ | $100_2$ | – | – | – | – |
| 19.6 | 345 | $-0.13_1$ | $0.26_1$ | $88_1$ | 2.8 | $-0.09_1$ | $0.29_1$ | $11_1$ |
| 19.7 | 347 | $-0.14_1$ | $0.33_1$ | $78_3$ | $2.4_1$ | $-0.09_1$ | $0.25_1$ | $22_1$ |
| 21.5 | 366 | $-0.14_1$ | $0.33_1$ | $66_4$ | $2.7_1$ | $-0.09_1$ | $0.25_1$ | $34_1$ |
| 25.4 | 398 | $-0.13_1$ | $0.26_1$ | $44_2$ | $2.9_1$ | $-0.09_1$ | $0.29_1$ | $56_1$ |
| 27.6 | 412 | $-0.13_1$ | $0.26_1$ | $35_3$ | $2.8_1$ | $-0.09_1$ | $0.26_1$ | $65_1$ |

The ordering temperatures, $T_N^\gamma$ are also included.

ND results. Then, the spectra were analyzed considering a non-resolved magnetic interaction representing antiferromagnetic $\gamma$ phase ($I_1$) and a single line representing paramagnetic $\varepsilon$ ($I_2$) phase. The resulting hyperfine parameters and relative fractions are quoted in Table I together with the corresponding $T_N^\gamma$ taken from literature [13].

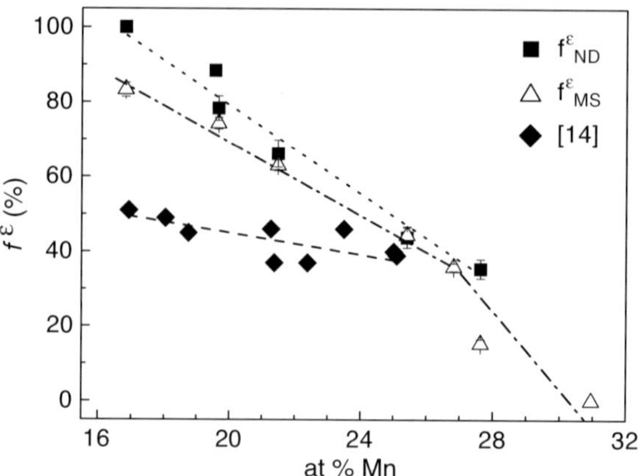


*Figure 4.* Variation with Mn content of the $\varepsilon$ phase fraction obtained from Mössbauer spectroscopy $\left(f^{\varepsilon}_{\mathrm{MS}}\right)$ and neutron diffraction $\left(f^{\varepsilon}_{\mathrm{ND}}\right)$ experiments. Results of dilatometry extracted from ref. [14] are included for comparison.

Finally, Figure 4 shows the present $f^{\varepsilon}_{\mathrm{MS}}$ and $f^{\varepsilon}_{\mathrm{ND}}$ and those extracted from literature [14] corresponding to dilatometric measurements. It is observed that the MS and ND determinations present similar trends, and noticeably differ at the low concentration range. This difference could be related to a higher density of stacking faults (SFs) present in the $\gamma$-phase, since at the 16 at.% Mn composition the SF energy has a minimum [15]. Due to the local atomic order, these SFs generate $\varepsilon$-like environments. Then, MS could be overestimating the $\varepsilon$ relative fraction in these cases. The other possibility is the presence of $\gamma$ paramagnetic phase that also could be shared in the $\varepsilon$ MS signal. Figure 4 also shows that $f^{\varepsilon}_{\mathrm{MS}}$ and $f^{\varepsilon}_{\mathrm{ND}}$ are systematically higher than those obtained from dilatometry [14]. This discrepancy could be associated with the self-accommodation of variants [16], or may be related to the polycrystallinity of the samples, giving then a macroscopic volume change, which does not necessarily reflects the real $\varepsilon$ phase fraction determined by dilatometry.

## 4. Conclusions

The evolution with Mn concentration of the $\varepsilon$ relative fractions obtained by MS and ND is similar, decreasing with the Mn content. The resulting MS and ND fractions are always larger than those obtained from dilatometry suggesting that the last technique becomes inappropriate to deduce phase fractions in the present situation. However, the use of MS for phase fraction detection in the Fe–Mn system where both phases $\varepsilon$ and paramagnetic $\gamma$ have very similar hyperfine parameters can be difficult, even more when a high concentration of SFs is expected. An additional technique like ND was necessary to use to clarify such a situation.

## Acknowledgements

We wish to thank the support of the Spanish Cooperation Research Group at the Institut Laue–Langevin, which allowed us to use the D1B neutron diffractometer. Research grants from CONICET (Consejo Nacional de Investigaciones Científicas y Técnicas, Argentina), Fundación Rocca and Fundación Antorchas from Argentina are gratefully acknowledged.

## References

1. Kopitsa G. P., Runov V. V., Grigoriev S. V., Bliznuk V. V., Gavriljuk V. G. and Glavatska N. I., *Physica B* **335** (2003), 134.
2. Cotes S., Fernández Guillermet A. and Sade M., *Mater. Sci.Eng., A* **273–275** (1999), 503.
3. Cotes S., Sade y M. and Fernández Guillermet A., *Met. and Mat. Trans. A*, **26A** (1995), 1957.
4. Martínez J., Cotes S. M., Cabrera A. F., Desimoni J. and Fernández Guillermet A., *Met. Trans., A*, send to publication in march 2004.
5. Benitez G. M., Aurelio G., Fernández Guillermet A., Cuello G. J. and Bermejo F. J., *J. Alloys Comp.* **284** (1999), 251.
6. Young R. A. (ed.), *The Rietveld Method*, Oxford University Press, New York, 1995.
7. Rodríguez-Carvajal J., Fullprof version 35d, LLB-JRC, France, 1998.
8. Umebashashi H. and Ishikawa Y., *J. Phys. Soc. Jpn.* **21**(7) (1966), 1281.
9. Hirai K. and Jo T., *J. Phys. Soc. Jpn.* **54**(9) (1985), 3567.
10. Schulthess T. C., Butler W. H., Stocks G. M., Maat S. and Mankey G. J., *J. Appl. Phys.* **85**(8) (1999), 4842–4844.
11. Marinelli P., Baruj A., Sade M. and Fernández Guillermet A., *Z. Met. Kd*, **91**(11) (2000), 957–962.
12. Marinelli P., Baruj A., Sade M. and Fernández Guillermet A., *Z. Met. Kd.* **92**(5) (2001), 489–493.
13. Huang W., *CALPHAD* **13** (1989), 243–252.
14. Marinelli P., Baruj A., Pons J., Sade M., Fernández Guillermet A. and Cesari E., *Mater. Sci. Eng., A.* **335** (2002), 137–146.
15. Cotes S. M., Fernández Guillermet A. and Sade M., *Met. Trans., A* **35A** (2004), 83.
16. Yang J. H. and Wayman C. M., *Mat. Charact.* **28** (1992), 28–37.

Hyperfine Interactions (2005) 161:229–235
DOI 10.1007/s10751-005-9195-z


# Effects of Hole-Doping on Superconducting Properties in $MgCNi_3$ and its Relation to Magnetism

M. ALZAMORA*, D. R. SÁNCHEZ, M. CINDRA and E. M. BAGGIO-SAITOVITCH
*Centro Brasileiro de Pesquisas Físicas, Rua Xavier Sigaud 150, Urca, CEP 22290-180 Rio de Janeiro, Brazil*

**Abstract.** Low temperature Mössbauer experiments were performed in Fe-doped and in C-deficiency $MgCNi_3$. No magnetic moment was found for Fe in $MgC(Ni_{0.99}Fe_{0.01})_3$ sample and *no* magnetic hf field was observed at any temperature for *all* the samples. These results shown no evidence of magnetic fluctuation or magnetic ordering influencing the depress of superconductivity in hole-doped $MgCNi_3$.

**Key Words:** magnetism, MgCNi3, superconductivity.

## 1. Introduction

The discovery of superconductivity in $MgCNi_3$ ($T_c \sim 8$ K) [1] was very surprising due to its high content of Ni (82.9 at.%), element usually associated to magnetism. Cubic $MgCNi_3$ (*Pm3m*) forms a three-dimensional perovskite structure resembling the structure of the high-$T_c$ superconductors. However, in contrast to the high-$T_c$ perovskites this compound has no oxygen in its structure. The $MgCNi_3$ is especially important, because it may be considered as a bridge between high-$T_c$ cuprates and conventional intermetallic superconductors.

It was suggested that ferromagnetism and superconductivity may coexist due to the ferromagnetic instability caused by the high density of electronic states at the Fermi level $N(E_F)$ [2]. This material is close to a magnetic instability on hole doping (e.g., substituting Ni by Fe, Co, Mn or C deficiency). Early band structure calculations suggested the involving of magnetic interaction in the suppression of superconductivity in hole-doped $MgCNi_3$ [3]. In fact, hole doping quickly destroy the superconductivity in $MgCNi_3$ [1, 4, 5]. Some calculations [5–7] showed that hole doping of $MgCNi_3$ is accompanied by a reduction of the density of states at the Fermi level, which seems to be responsible for the reduced superconductivity, but magnetic ground states are only expected for high hole dopants (Fe, Co) [6] or absolute absence of C [8].

* Corresponding author. Fax: + 55- 21- 21417540. E- mail: mariella@cbpf.br

In this paper we report X-ray diffraction, magnetization and Mössbauer measurements, down to 1.5 K, in hole-doped polycrystalline powder of $MgC(Ni_{1-x}Fe_x)_3$ ($0.01 < x < 0.1$) and $MgC_{0.8}(Ni_{0.99}Fe_{0.01})_3$. We investigate the hole-doping effect on superconducting and magnetic properties in Fe doped $MgCyNi_3$. The Fe can be used as well as a hole dopant or local probe at the Ni site. In spite $^{57}Fe$ Mössbauer spectroscopy is not the appropriate technique to study superconducting properties, it is a powerful tool to study magnetism. Information about the negative role of magnetism in the suppression of superconductivity in hole-doped $MgCNi_3$ was obtained.

## 2. Experimental

The preparation of the $MgC_y(Ni_{1-x}Fe_x)_3$ ($x = 0.01, 0.02, 0.1$ and $y = 0.8, 1.0$) samples is described elsewhere [1, 4]. We used $^{57}Fe$ to prepare the samples $MgC_y(Ni_{0.99}Fe_{0.01})_3$ ($y = 0.8, 1.0$). The structural characterizations were made by powder X-ray diffraction using a Rigaku X-ray diffractometer with $CuK_\alpha$ radiation. The samples were analyzed by Mössbauer spectroscopy in transmission geometry in the temperature range of 1.5–300 K and in an external applied magnetic field up to 7 T. AC susceptibility measurements were performed in a SQUID magnetometer.

## 3. Results and discussion

X-ray characterization indicates the single cubic $MgCNi_3$ as the majority phase for all the samples (Figure 1). A small amount of unreacted graphite was found for the samples $MgC(Ni_{1-x}Fe_x)_3$ due to the excess of C utilized for its synthesis [1, 4].

The refined lattice parameter $a$ increase (decrease) as the Fe doping (C deficiency) increase respect to that of the undoped $MgCNi_3$. The lattice parameter $a = 3.79$ Å obtained for $MgC_{0.8}(Ni_{0.99}Fe_{0.01})_3$ is consistent with that found for nonsuperconducting $MgC_{0.8}Ni_3$ in previous reports [5, 9]. For the samples $MgC(Ni_{1-x}Fe_x)_3$ ($x = 0.01$ and $0.02$) the lattice parameters $a$ are almost the same as that for the undoped $MgCNi_3$ ($a = 3.8118$ Å) compound, while for the sample with $x = 0.1$ a slightly increase was observed ($a = 3.8174$ Å).

The ac susceptibility measurements, showing the superconducting transition of the doped materials, are displayed in Figure 2. The Fe dilution at the Ni site quickly depress the $T_c$ from $\sim 7$ K for $x = 0.0$ to $T_c \sim 3$ K for $x = 0.02$, and the disappearance of the superconducting state is expected to occurs at $x \sim 0.04$ (inset Figure 2). No diamagnetic signal is observed for the compound with $x = 0.1$.

In order to detect any possible magnetic interaction influence on the superconducting depression, Mössbauer spectroscopy as a function of temperature

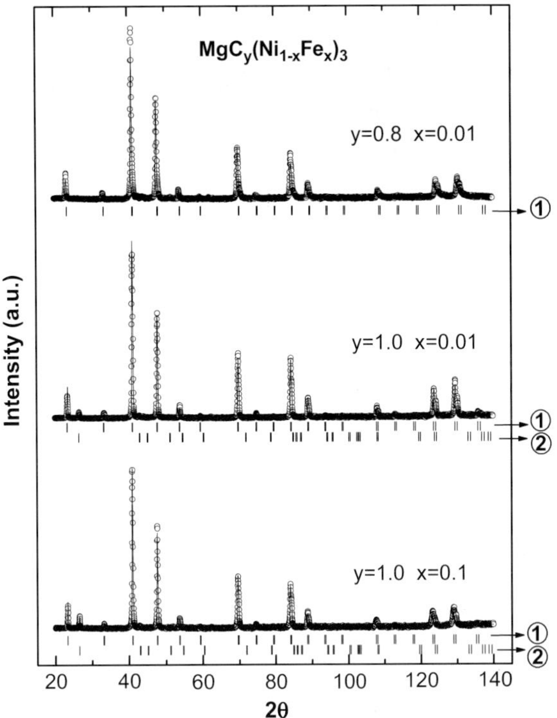


*Figure 1.* Cu − $K_\alpha$ X-ray diffraction pattern of the samples $MgC_{0.8}(Ni_{0.99}Fe_{0.01})_3$ and MgC $(Ni_{1-x}Fe_x)_3$ ($x$ = 0.01, 0.1). The Bragg peaks, corresponding to the $MagCNi_3$ (*1 vertical line*) and graphite (*2*), obtained after Rietveld analysis, are indicated.

was performed. The room temperature spectra show a single quadrupole doublet (Figure 3a), which is attributed to Fe in the $MgCNi_3$ structure. The sharp resonance lines indicate a single well-defined site for Fe in these compounds. We want to mention that a weak doublet caused by an impurity phase and a weak sextet (unreacted Fe) were taken into account in the analysis of all the low-temperature $^{57}Fe$ ME spectra shown below.

Comparing the 4.2 K spectra (Figure 3b) with that at RT (Figure 3a) there is no obvious difference: The value of the isomer shift is slightly more negative compared to the center shift at RT due to second-order Doppler shift, the line-width is the same within the experimental errors, only the electrical quadrupole splitting $\Delta E_Q$, a measure of the local structure is slightly smaller. Additional spectra were taken at temperatures between 300 and 4.2 K. Neutron diffraction studies [10] have shown a slight thermal contraction in $MgCNi_3$ compound as the temperature decrease: the lattice contracts down to ~20 K and for lower temperatures, where superconductivity appears, it remains almost constant. The $\Delta E_Q$ values for all the samples studied here decrease slightly with temperature decreasing (Figure 4), as expected on the basis of a point-ion lattice field varying with thermal contraction of the lattice. The variation of $\Delta E_Q$ with temperature

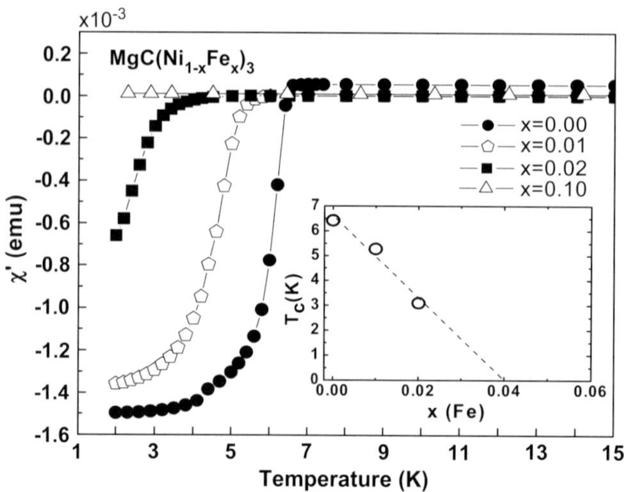


*Figure 2.* AC susceptibility curves for $MgC(Ni_{1-x}Fe_x)_3$ ($x = 0.01, 0.02, 0.1$). Inset: variation of $T_c$ with Fe concentration.

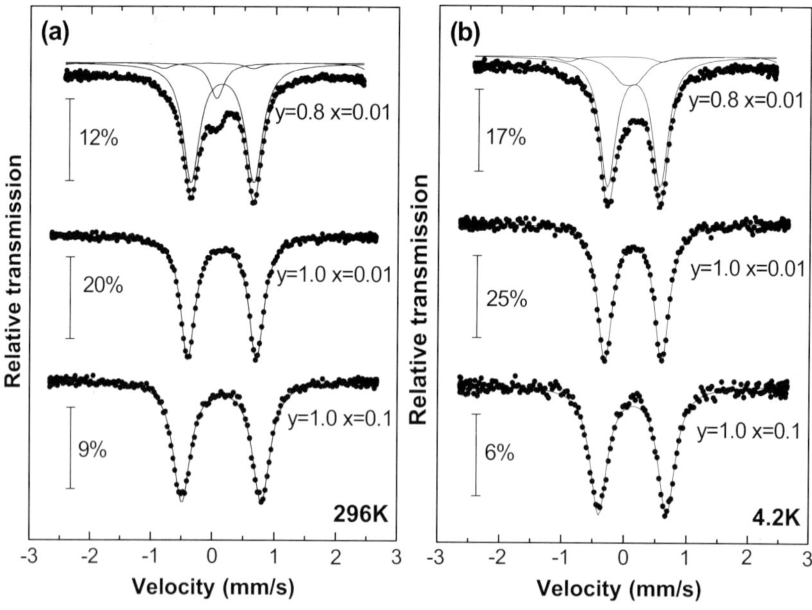


*Figure 3.* (a) Room temperature and (b) 4.2 K $^{57}Fe$ Mössbauer spectra of $MgC_{0.8}(Ni_{0.99}Fe_{0.01})_3$ and $MgC(Ni_{1-x}Fe_x)_3$ ($x = 0.01, 0.1$).

does not show structural changes between 4.2 and 300 K, only resembles the temperature dependence of the lattice parameter *a* [10].

The variation of $|\Delta E_Q|$ values from 300 to 4.2 K for $MgC_{0.8}(Ni_{0.99}Fe_{0.01})_3$ and $MgC(Ni_{0.99}Fe_{0.01})_3$ is ~21.5% (Figure 4). The $|\Delta E_Q|$ of 1.08 mm/s obtained for $MgC(Ni_{0.9}Fe_{0.1})_3$ at 4.2 K correspond to a change of 19% respect to its

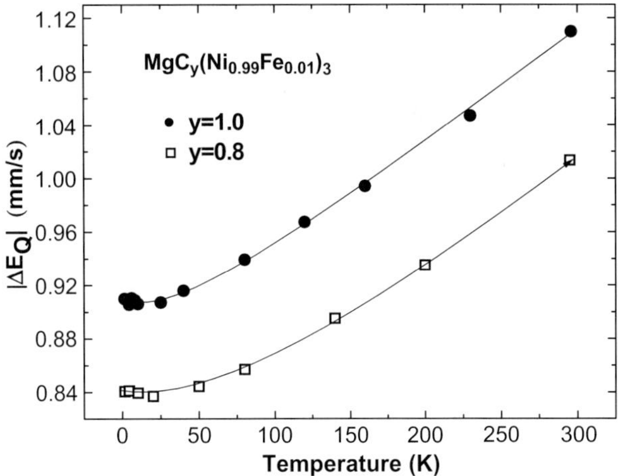


*Figure 4.* $|\Delta E_Q|$ for $MgC_{0.8}(Ni_{0.99}Fe_{0.01})_3$ and $MgC(Ni_{0.99}Fe_{0.01})_3$ as a function of temperature.

value at 300 K (1.28 mm/s), an expected value if we take into account the change observed for the other samples. So, justified the variation of $\Delta E_Q$ with temperature, we can conclude that *no* magnetic hf field was observed at any temperature for *all* the samples.

To get information on magnetic moment of Fe in these compound we performed low-temperature Mössbauer measurements under external applied magnetic field. The 4.2 K Mössbauer spectra under different external applied magnetic field of the $MgC(Ni_{0.99}Fe_{0.01})_3$ sample are shown in Figure 5. The magnetic moments governing the local magnetization can be derived from the measured internal fields $B_{eff} = B_{hf} + B_{ext}$. All the spectra were analyzed with a least-squares fitting yielding to, within the experimental error, $B_{eff} = B_{ext}$ (i.e., $\mu_{Fe} < 0.01\mu_B$) indicating no magnetic moment for Fe. Furthermore, it was found that $V_{ZZ} < 0$ indicating a predominant contribution of $d_z^2$, $d_{xz}$ and $d_{yz}$ orbitals to $\Delta E_Q$.

Being the highly hole-doped $MgC_{0.8}(Ni_{0.99}Fe_{0.01})_3$ and $MgC(Ni_{0.9}Fe_{0.1})_3$ not superconducting and taking into account that no signal of short or long magnetic ordering nor magnetic fluctuation was detected, we can conclude that hole doping does not lead the system to a magnetic state which could be responsible for pair breaking. These results are in contradiction with early band structure calculations suggesting the involving of magnetic interaction in the suppression of superconductivity in hole-doped $MgCNi_3$ [3], but in agreement with recent results indicating no evidence for long-range magnetic ordering [5, 9] but a decrease of $N(E_F)$ [5, 6]. So, Fe dopant in $MgCNi_3$ behaves as a source of *d*-bad holes rather than magnetic scattering centers, broadening the band width and shifting the location of $E_F$ yielding to a reduction of $N(E_F)$ [6]. The carbon-site vacancies in $MgCNi_3$ may enhance the itinerancy of electrons leading to a depress of the DOS peak below $E_F$ and a consequent reduction of $N(E_F)$ [5]. On

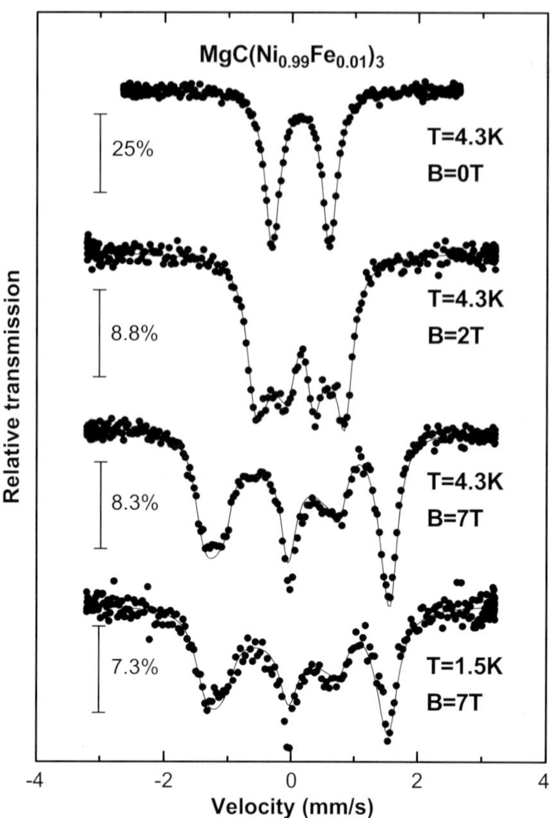


*Figure 5.* The 4.2 and 1.5 K $^{57}Fe$ Mössbauer spectra of $MgC(Ni_{0.99}Fe_{0.01})_3$ under different external applied magnetic field (B).

the other hand, neutron studies [9] have shown that the introduction of carbon vacancies has a significant effect on the positions of the Ni atoms. Structural effects on the electron–phonon coupling may play an important role in describing the nature of the depression of $T_c$. Hence, the structural effects, due to the introduction of hole dopants, should be included in the analysis of the ground state properties of these compounds.

## 4. Conclusion

In summary, we have investigated the X-ray, ac susceptibility and low-temperature Mössbauer data in Fe doped and in C deficiency $MgCNi_3$. We found that Fe has no magnetic moment and no magnetic hf field at the Fe nucleus at any temperature was observed for all the samples, showing a clear evidence for absence of magnetic fluctuation or magnetic ordering and ruling out the influence of magnetism in the depression of superconductivity of hole-doped $MgCNi_3$ compound.

## References


1. He T., Huang Q., Ramirez A. P., Wang Y., Regan K. A., Rogado N., Hayward M. A., Haas M. K., Slusky J. S., Inumara K., Zandbergen H. W., Ong N. P. and Cava R. J., *Nature (London)* **411** (2001), 54.
2. Rosner H., Weht R., Johannes M. D., Pickett W. E. and Tosatti E., *Phys. Rev. Lett.* **88** (2002), 027001.
3. Wang J. L., Yu X., Zeng Z., Zheng Q. Q. and Lin H. Q., *J. Appl. Phys.* **91** (2002), 8504.
4. Alzamora M., Sánchez D. R., Cindra M. and Baggio-Saitovitch E. M., *Braz. J. Phys.* **32** (2002), 755.
5. Shan L., Xia K., Liu Z. Y., Wen H. H., Ren Z. A., Che G. C. and Zhao Z. X., *Phys. Rev. B* **68** (2003), 024523.
6. Kim I. G., Lee J. I. and Freeman A. J., *Phys. Rev. B* **65** (2002), 064525.
7. Shein I. R., Ivanovskii A. L., Kurmaev E. Z., Moewes A., Chiuzbian S., Finkelstein L. D., Neumann M., Ren Z. A. and Che G. C., *Phys. Rev. B* **66** (2002), 024520.
8. Shim J. H., Kwon S. K. and Min B. I., *Phys. Rev. B* **64** (2001), 180510.
9. Amos T. G., Huang Q., Lynn J. W., He T. and Cava R. J., *Solid State Commun.* **121** (2002), 73.
10. Huang Q., He T., Regan K. A., Rogado N., Hayward M., Haas M. K., Inumaru K. and Cava R. J., *Physica C* **363** (2001), 215.

Hyperfine Interactions (2005) 161:237–246
DOI 10.1007/s10751-005-9196-y 

# Influence of Electron Delocalization on the Magnetic Properties of Iron Ludwigite $Fe_3O_2BO_3$

J. J. LARREA[1,*], D. R. SANCHEZ[1], F. J. LITTERST[1,2] and E. M. BAGGIO-SAITOVITCH[1]
[1]*Centro Brasileiro de Pesquisas Físicas, Rua Dr. Xavier Sigaud 150, Rio de Janeiro, 22290-180, RJ, Brazil; e-mail: jlarreaj@cbpf.br*
[2]*Institute for Metal Physics and Nuclear Solid State Physics, Technische Universität Braunschweig, Mendelssohnstr. 3, D 38106, Braunschweig, Germany*

**Abstract.** $^{57}$Fe Mössbauer spectroscopic studies of iron ludwigite $Fe_3O_2BO_3$ performed between 4 and 450 K allow the discussion of magnetic spin arrangements and the dynamics of electronic configurations of iron. The observed magnetic transitions are related to charge ordering.

**Key Words:** antiferromagnetism, charge order, double exchange, iron oxyborate, Mössbauer spectroscopy, weak ferromagnetism.

## 1. Introduction

The homometallic iron ludwigite $Fe_3O_2BO_3$ is an oxyborate with orthorhombic structure (see Figure 1) [1–6]. It is basically formed from two subsystems. Firstly, there are three leg ladders (3LL) directed along the *c* axis (i.e., perpendicular to the plane shown in Figure 1), they are consisting of oxygen octahedra with iron in its centres. The rungs of the 3LL are formed by three iron sites (one site 2 and two sites 3) forming a so-called triad. Secondly, there are oxygen octahedra with iron site 4 and site 1, which together with the boron atoms (open circles in Figure 1) are connecting the 3LL. Neighbouring 3LL together with sites 1 and 4 between them can be conceived as zigzag walls extending along *c*.

Sites 1 and 4 are magnetically and also from charge state clearly distinct from sites 2 and 3 of the 3LL which are mainly determining the properties of the iron ludwigite. Reason is the mixed valent character of the iron ions in sites 2 and 3 in contrast to those in sites 1 and 4, which are clearly high-spin $Fe^{2+}$.

From formal stoichiometry considerations each triad can be considered as being formed by three $Fe^{3+}$ with an extra electron added. These extra electrons

* Author for correspondence.

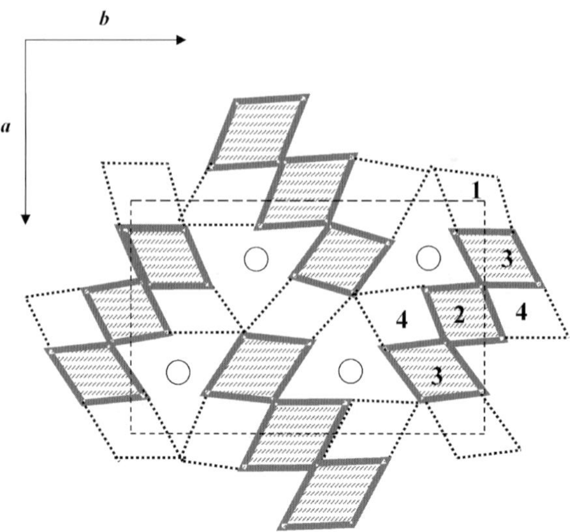


*Figure 1.* Projection of $Fe_3O_2BO_3$ structure to *ab* plane. Iron sites 1 to 4 are in the centres of oxygen octahedra. The 3-2-3 marks the triads, which are the rungs of the 3LL running along the *c* direction (i.e. out of plane). *Circles* indicate boron.

are distributed along the 3LL and are strongly correlated. They are supposed to cause peculiar charge and spin dynamics. Macroscopic thermodynamic and transport properties have been studied [5] giving evidence for semiconducting behaviour with an activation energy corresponding to 60 and 1300 K for the temperature ranges below and above about 220 K, respectively, the supposed onset temperature for charge delocalization along the 3LL. Caloric data can be interpreted with 2d magnon excitations below about 40 K. Between about 110 and 200 K a linear specific heat is interpreted with excitations in a Wigner glass caused by tunnelling of the extra electrons.

Neutron diffraction indicates an antiferromagnetic state (called AF2) for the 3LL at 5 K with an up–down–up structure at sites 3, 2 and 3 of the triads [7] with an overall ferromagnetic behaviour which is in contradiction to magnetization which indicates an antiferromagnetic ground state [5]. Magnetization data show a weak ferromagnetic phase (WF) between about 50 and 74 K [5]. Mössbauer spectroscopy allows distinguishing in addition a third magnetically ordered phase between 74 and 112 K [8]. From magnetization it becomes clear that this is a second antiferromagnetic phase (AF1) [5].

Earlier Mössbauer results [4] have given evidence for the presence of mixed valent iron. We extended these studies [2] and detected a charge order transition at $T_{CO} \approx 300$ K, which afterwards also could be detected from X-ray diffraction [3]. The changes of charge states at lower temperatures and the magnetic behaviour are described in ref. [9]. Our data were mainly corroborated by the

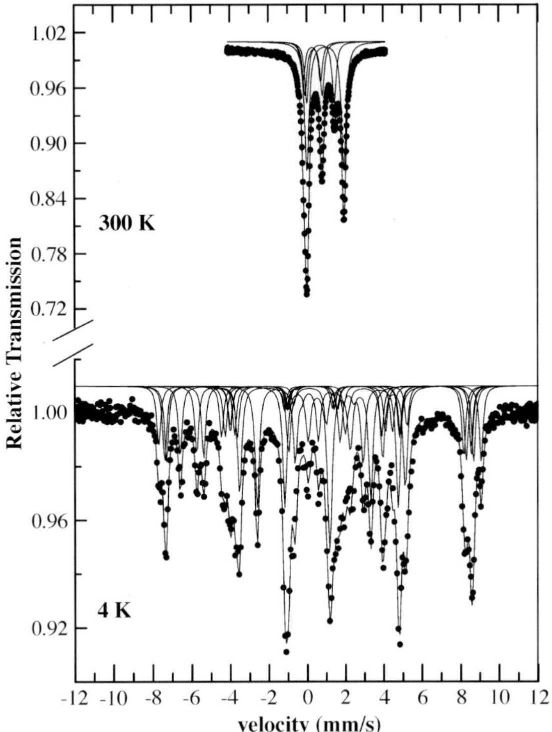


*Figure 2.* Typical $^{57}$Fe Mössbauer absorption spectra in the paramagnetic (*upper*) and the magnetically ordered AF2 phase (*lower*).

studies of Douvalis et al. [10]. In contrast to our results their data cannot distinguish the transition between the low temperature AF2 and the WF phase which may be a problem of different data analysis [9].

Several interesting theoretical considerations concerning charge order and delocalization and the possible influence of the dimensionality have been presented [6, 11, 12]. In this present paper we want to give a critical résumé of Mössbauer data in relation with these models since we believe that hyperfine data provide essential information about the individual iron sites' charge states and their magnetic properties. Since sites 1 and 4 do not participate in charge dynamics and in charge order we will concentrate here on the properties of the 3LL.

## 2. Summary of Mössbauer spectroscopic results

We only want to give here a brief overview of the Mössbauer data. Detailed information is given elsewhere [9]. Typical spectra for the paramagnetic state and the low temperature ordered state are shown in Figure 2. The complex spectra

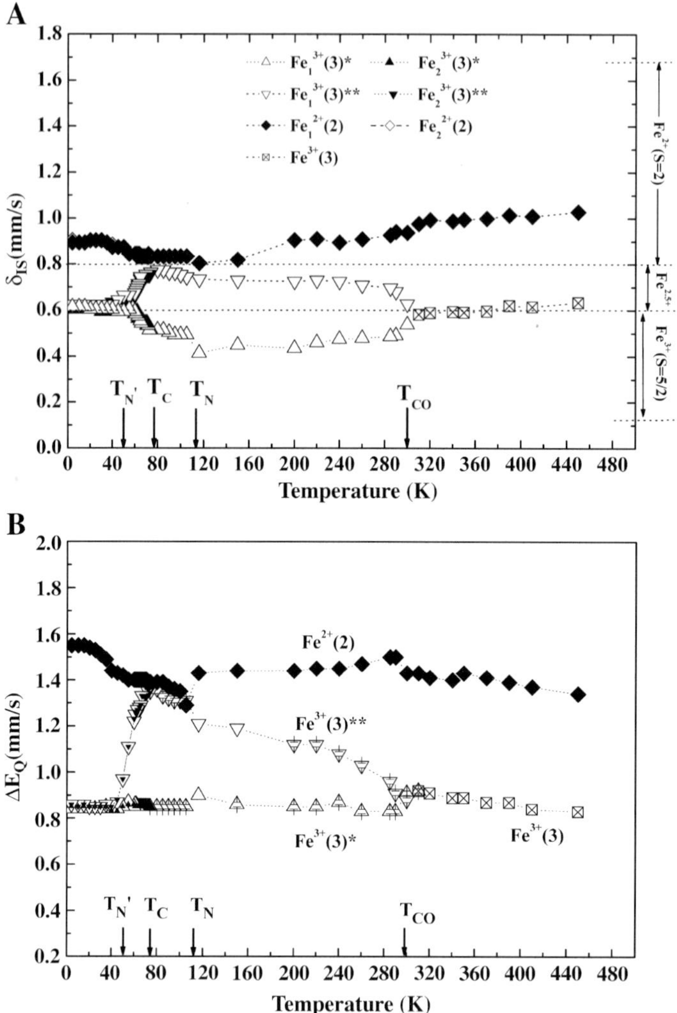


*Figure 3.* (A) Temperature dependence of isomer shifts of the triad sites. (B) Temperature dependence of the quadrupole splitting of the triad sites.

can be analyzed with the superposition of subspectra for individual sites 1 to 4. Isomer shifts and quadrupole splittings for sites 1 and 4 are typical for divalent iron without exceptional changes. For the triad sites, however, these parameters are revealing remarkable variations (Figure 3A and B). Above $T_{CO} \approx 300$ K sites 3 ($Fe^{3+}$(3)) have the same isomer shifts and quadrupole splittings indicating a charge state, which is not clearly trivalent but rather mixed valent; $Fe^{2+}$(2) is divalent. Below $T_{CO}$ sites 3 reveal two different isomer shifts and quadrupole interactions. We call these sites $Fe^{3+}$(3)* and $Fe^{3+}$(3)**, the latter having even more mixed valent character than $Fe^{3+}$(3), the other, however, is more trivalent

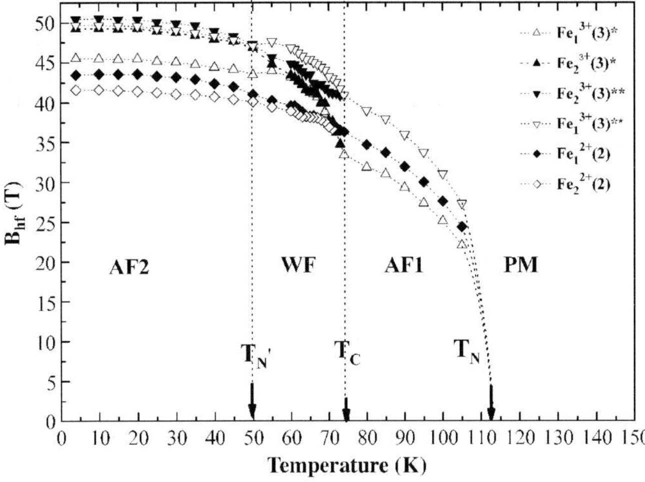


*Figure 4.* Temperature dependence of the magnetic hyperfine fields of the triad sites.

(see isomer shifts). Between about 120 and 75 K (the range of the AF1 phase) $Fe^{3+}(3)^{**}$ and $Fe^{2+}(2)$ can hardly be distinguished from isomer shift and quadrupole interaction. Below 74 K (in the WF phase) $Fe^{3+}(3)^{**}$ and $Fe^{3+}(3)^*$ approach again in their charge states and below about 50 K (AF2 phase) both are close to the state found above $T_{CO}$. Site 2 then is divalent, yet still close to mixed valent character. The magnetic hyperfine fields for the triads are shown in Figure 4. They are revealing an increase of sublattice magnetizations below $T_N = 112$ K with a further increase below $T_C = 74$ K in the WF phase. In the WF phase a further complication appears: the spectra can only be fitted assuming two magnetically different subsites for each site revealing different magnetic hyperfine fields yet the same isomer shifts and quadrupole interactions. The same holds below $T_N' = 50$ K in AF2, there however with less deviations from the average for each site. Again this is in severe contrast to sites 1 and 4: whereas sites 4 only show a very small field in AF2, probably only of transferred nature, sites 1 reveal an onset of magnetic hyperfine field only below $T_C$. Below $T_C$ the magnetic hyperfine fields for both sites show typical magnetization behaviour, however, with clearly smaller saturation fields compared to those of the triad sites. This indicates ordering of the 3LL below $T_N$, sites 1 and 4 which are between the 3LL, however, are still magnetically decoupled and order only at $T_C$. All sites exhibit considerable canting of the magnetic hyperfine field with respect to the main axis of the electrical field gradient. The canting angles are changing considerably when passing $T_N'$ and $T_C$. We only want to give here some qualitative arguments since an unambiguous data analysis deriving both polar and azimuthal angles together with an eventually non-zero asymmetry parameter of the electrical field gradient is not possible for a nuclear spin 3/2 to 1/2 transition [13].

## 3. Discussion

We now will discuss the various regimes with special respect to the information we draw from the hyperfine data: up to $T_N'$; between $T_N'$ and $T_C$; between $T_C$ and $T_N$; between $T_N$ and $T_{CO}$; above $T_{CO}$.

### 3.1. $T < T_N'$

Isomer shifts and quadrupole splittings of the triad sites in the AF2 phase indicate that sites 3 are identical from charge point of view, but they have different magnetic hyperfine fields and different canting. The isomer shift values of sites 2 and 3 indicate that the extra electron is mainly located at site 2, i.e., in the middle of the 3LL. From electrostatic consideration within a single 3LL one might expect to have the extra electron charge mainly located in the outer ladder. In fact the hyperfine parameters of sites 3 indicate that there is a non-negligible charge transfer to sites 3. This is also expected from the relatively short distances between sites 3 and 2 of about 2.79 Å with some overlap of their 3d wave functions allowing charge hopping at least at higher temperatures. One should, however, take into account also the charges outside the 3LL, namely the divalent ions in sites 1 and 4 favouring the localization of the extra charge in sites 2.

The distribution of values of magnetic hyperfine fields and also of angles points to a severely canted and probably modulated antiferromagnetic structure. This is not unexpected due to the just described overlap of 3d wave functions. As shown by de Gennes [14] under these conditions double exchange may lead to considerably canted, helical, or even disordered moment arrangements in the ground state. From our data a possible structure would be a probably incommensurate spiral structure for all three legs of the 3LL. The gross arrangement in one triad is up–down–up (in agreement with neutron results) or inverse, yet with considerable canting. Further reasons for canting are the different anisotropies due to the 3d orbitals in different symmetries of the iron states and the non-negligible interaction between neighbouring 3LL which is mediated via sites 1 and 4. A schematic view of charge distribution in the 3LL is shown in Figure 5. The charge distribution is ordered and at the time scale of Mössbauer spectroscopy (about microsecond) static. A possible moment arrangement is given in Figure 6 where the projection to the plane of a single 3LL is shown indicating also whether the moments are pointing into or out of this plane.

### 3.2. $T_N' < T < T_C$

In the WF regime one observes a gradual change of the charge distribution: one of sites 3 is turning more trivalent ($Fe^{3+}(3)^*$), the other tends to accept 3d charge coming closer with its hyperfine parameters to those of $Fe^{2+}(2)$. This shift of charge mainly from one side of the triad to the other marks a transition between AF2 and AF1 where $Fe^{3+}(3)^*$ and $Fe^{2+}(2)$ become indistinguishable from charge

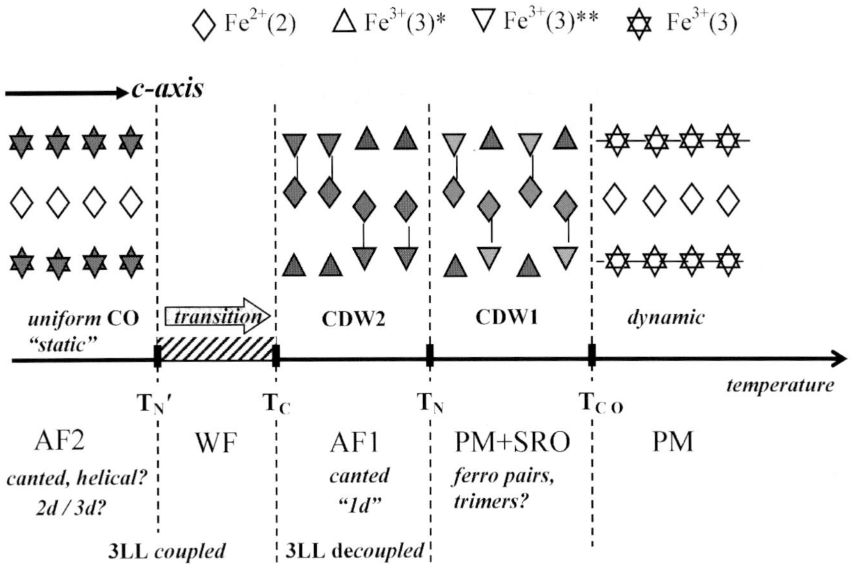


*Figure 5.* Proposed charge (*upper part*) and magnetic arrangement (*lower part*) of the 3LL running along *c*.

state. The magnetic patterns in the WF range reveal an even more complex arrangement than in AF2 mainly with still stronger canting and deviations of magnetic hyperfine fields from the average ones [9]. We cannot exclude that one has to deal with a mixture of AF1 and AF2 leading locally to frustration.

Another reason for the staggered magnetization of the WF could be an exchange mechanism of Dzialoshinski–Moriya type [15].

### 3.3. $T_C < T < T_N$

As mentioned above there is a clear formation of pairs in AF1 with equal mixed valent charge state and also symmetry as seen from quadrupole interaction whereas one site 3 is clearly trivalent. The major change concerning magnetic behaviour is that sites 1 and 4 turn non-(or nearly non-)magnetic, i.e., the coupling between neighbouring 3LL is vanishing. Each triad site now reveals one magnetic hyperfine field only. The temperature dependences of the hyperfine fields follow magnetization curves with clearly lower (extrapolated) saturation fields than for WF and AF2. In this situation with decoupled 3LL one can expect that the one- dimensional character of the ladders becomes pronounced and one might suspect the formation of non-magnetic singlets [9]. For this, however, no evidence is found. Reason may be the strongly correlated behaviour due to the mixed valent ions. We propose that the mixed valent pairs couple mainly ferromagnetically as expected from double exchange, with the second trivalent ion

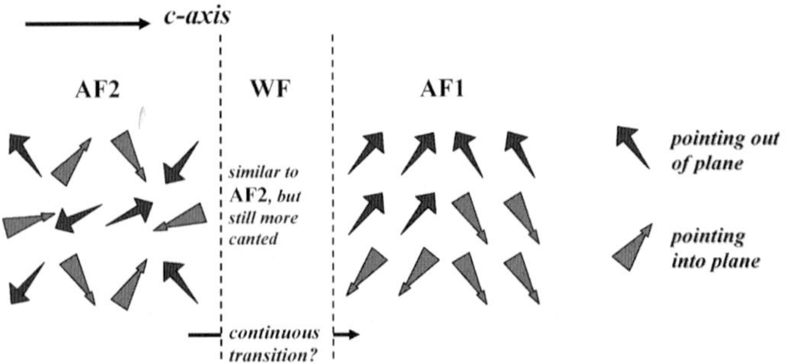


*Figure 6.* Proposed canted magnetic structures of 3LL.

in site 3 coupling antiparallel to the pair. A possible charge arrangement is given by a charge density wave (CDW2, see Figure 5). The moment orientation is still canted with main direction perpendicular to *c* close to the normal of the plane formed by each 3LL. A proposed magnetic structure is given in Figure 6.

### 3.4. $T_N < T < T_{CO}$

The magnetic hyperfine interaction has vanished and the charge distribution in the triads gradually relaxes, meaning that the two ions of the mixed valent pairs become again distinguishable. We believe that the structure is best characterized now by a different charge density wave (CDW1) for which evidence comes from X-ray diffraction [3].

Also in this regime there is no evidence for non-magnetic singlets from trivalent iron as suspected in theoretical considerations together with magnetization data yielding relatively small effective moments. Our Mössbauer spectra in applied magnetic field rather indicate short-range order (SRO) with moments around 10–15 $\mu_B$ from which one may suspect ferromagnetic dimers or trimers embedded in a paramagnetic (PM) surrounding. This finding is also compatible with the magnetization data. In contrast to [10] who assumed quenched orbital moments for all divalent iron ions we assume that the orbital moments of the divalent ions in the triad are quenched and also those in sites 1 and 4 are reduced. This is supported by the values of magnetic hyperfine fields.

Close to $T_{CO}$ the hyperfine parameters of sites 3 approach each other indicating the transition from the CDW1 to a more uniform distribution.

### 3.5. $T_{CO} < T$

This is reached at $T_{CO}$ where sites 3 become again indistinguishable from isomer shift and quadrupole interaction. The uniform charge distribution along sites 3 and sites 4, respectively, is reflected in the high temperature orthorhombic structure in contrast to the structure below $T_{CO}$, which possesses a doubled cell along *c*.

This transition was for the first time proposed from the Mössbauer data and indicates a fast charge delocalization along the c-direction of the 3LL.

## 4. Conclusions

As main results concerning charge order and magnetism in the 3LL of iron ludwigite and we can summarize:

– The extra 3d electrons of iron in the triads are mainly in sites 2, i.e., sites in the middle of the 3LL. We propose that this is due to the influence of $Fe^{2+}$ in sites 1 and 4.
– For all temperatures the magnetic properties of the 3LL are controlled by electron (de-)localization.
– From the spectra in applied field above 112 K there is no evidence for non-magnetic singlets of antiferromagnetically coupled pairs of $Fe^{3+}$. However, there are indications for short-range order with effective moments of about 10–15 $\mu_B$ what is compatible with magnetization data. The moments can be interpreted with ferromagnetically coupled dimers of $Fe^{3+}$(3)**–$Fe^{2+}$(2), or even trimers.
– Iron in sites 1 and 4 is not magnetically ordered down to 74 K, i.e., neighbouring 3LL are magnetically decoupled.
– The ordered magnetic structures below $T_N$ are complex. We propose that they are related to different charge arrangements in charge density waves and that double exchange plays an important role for the details.
– The origin of the weak ferromagnetic phase remains unclear, also whether it is a magnetically and electronically homogeneous phase or a mixture of AF1 and AF2.
– Dimensionality does not primarily determine the magnetic properties of the ladders.
– There are indications from the Mössbauer data for additional structural changes which may occur at $T_N$ and $T_N$ and are related to charge order.

## Acknowledgements

E.M. Baggio-Saitovitch and F.J. Litterst are grateful to DAAD/CAPES and DAAD /CNPq for traveling support. E.M. Baggio-Saitovitch acknowledges partial support by FAPERJ (Cientista do Nosso Estado) and MCT (PRONEX 66201998-9).

## References

1. Fernandes J. C., Guimarães R. B., Continentino M. A., Borges H. A., Valarelli J. V. and Lacerda A., *Phys. Rev., B* **50** (1994), 16754.
2. Larrea J., Sánchez D. R., Baggio-Saitovitch E. M. and Litterst F. J., *J. Phys., Condens. Matter* **13** (2001), L949.

3. Mir M., Fernandes J. C., Guimarães R. B., Continentino M. A., Doriguetto A. C., Mascarenhas Y. P., Ellena J., Castellano E. E., Freitas R. S. and Ghivelder L., *Phys. Rev. Lett.* **87** (2001), 147201.
4. Swinnea J. S. and Steinfink H., *Am. Mineral.* **68** (1983), 827.
5. Guimarães R. B., Mir M., Fernandes J. C., Continentino M. A., Borges H. A., Cernicchiaro G., Fontes M. B., Candela D. R. S. and Baggio-Saitovitch E., *Phys. Rev., B* **60** (1999), 6617.
6. Fernandes J. C., Guimarães R. B., Continentino M. A., Ghivelder L. and Freitas R. S., *Phys. Rev., B* **61** (2000), R850.
7. Attfield J. P., Clarke J. F. and Perkins D. A., *Physica B* **180** (1992), 581.
8. Larrea J. J., Sánchez D. R., Baggio-Saitovitch E., Fernandes J. C., Guimarães R. B., Continentino M. A. and Litterst F. J., *J. Magn. Magn. Mat.* **226** (2001), 1079.
9. Larrea J. J., Sánchez D. R., Litterst F. J., Baggio-Saitovitch E., Fernandes J. C., Guimarães R. B. and Continentino M. A., *Phys. Rev., B* **70** (2004), 174452.
10. Douvalis A. P., Moukarika A., Bakas T., Kallias G. and Papaefthymiou V., *J. Phys., Condens. Matter* **14** (2002), 3302.
11. Latgé A. and Continentino M. A., Phys. Rev., *B* **66** (2002), 94113.
12. Whangbo M.-H., Koo H. J., Dumas J. and Continentino M. A., *Inorg. Chem.* **41** (2002), 2193.
13. Van Dongen Torman J., Jannathan R. and Trooster J. M., *Hyperfine Interact.* **1** (1975), 135.
14. de Gennes P. G., *Phys. Rev.* **118** (1960), 141.
15. Nathans R., Pickart S. J., Alperin H. A. and Brown P. J., *Phys. Rev.* **136** (1964), A1641.

Hyperfine Interactions (2005) 161:247–248 

# Author Index to Volume 161 (2005)